AF412990

METAL CLUSTERS
IN CHEMISTRY

METAL CLUSTERS
IN CHEMISTRY

PROCEEDINGS OF
A ROYAL SOCIETY DISCUSSION MEETING
HELD ON 20 AND 21 MAY 1982

ORGANIZED AND EDITED BY
SIR JACK LEWIS, F.R.S., AND M. L. H. GREEN

LONDON
THE ROYAL SOCIETY
1983

Printed in Great Britain for the Royal Society
at the
University Press, Cambridge

ISBN 0 85403 204 5

First published in *Philosophical Transactions of the Royal Society of London,*
series A, volume 308 (no. 1501), pages 1–166.

Published by the Royal Society
6 Carlton House Terrace, London SW1Y 5AG

CONTENTS

CONTENTS

Phil. Trans. R. Soc. Lond. A **308**, 3 (1982)
Printed in Great Britain

Introductory remarks

BY SIR JACK LEWIS, F.R.S.
University Chemical Laboratory, Lensfield Road, Cambridge CB2 1EW, U.K.

It gives me great pleasure to introduce what I believe is the first Discussion Meeting on inorganic chemistry organized by the Royal Society in the immediate past. In choosing the topic for this discussion it was generally felt that one of the most rapidly developing areas of inorganic chemistry is that of cluster compounds. This area of chemistry covers virtually the whole of the periodic table and today's discussion is concerned primarily with the cluster compounds formed by metal carbonyls, with illustrations of other areas of cluster chemistry of the sulphur and halide complexes of transition metals.

Cluster chemistry very much dominates the chemistry of the early transition series and is particularly significant with the second and third row elements. A study of the chemistry of cluster systems has emphasized that our initial ideas on the chemistry of mononuclear species must be considerably modified when considering the behaviour of polynuclear systems. Thus the bonding of a ligand in most cases occurs to more than one metal centre, and ligand bonding modes not found in mononuclear complexes are often experienced in the cluster complexes. Those results have obvious implications both in structural and reactivity patterns. In the papers that follow it is hoped to develop both these patterns in the polynuclear compounds and emphasize the contrasting nature of these to those of the well documented mononuclear series.

I have particularly avoided any attempt to define the term 'cluster'. I imagine that this will vary from speaker to speaker but the common theme of a multi-metal atom system should provide an interesting correlation between the various chemistries to be discussed.

Phil. Trans. R. Soc. Lond. A **308**, 5–15 (1982) [5]
Printed in Great Britain

Aspects of ruthenium and osmium cluster chemistry

By B. F. G. Johnson and Sir Jack Lewis, F.R.S.

University Chemical Laboratory, Lensfield Road, Cambridge CB2 1EW, U.K.

In this paper, we survey the synthesis, structure and bonding of a series of cluster carbonyl compounds of ruthenium and osmium containing from three to ten metal atoms. Members of this series include neutral, anionic hydrido and carbido clusters, all of which are derived from their $[M_3(CO)_{12}]$ parent. In general, the synthesis of these compounds involves the pyrolysis of either $[M_3(CO)_{12}]$ or, with osmium, some higher nuclearity cluster. The mechanism by which these reactions occur is believed to involve the formation of highly unstable, unsaturated derivatives. Many of these clusters contain frameworks of metal atoms that may be regarded as fragments of close-packed metallic arrangements; others may be regarded as examples of tetrahedral growth patterns. They exhibit a new and diverse chemistry, much of which may now be understood in terms of simple bonding arguments.

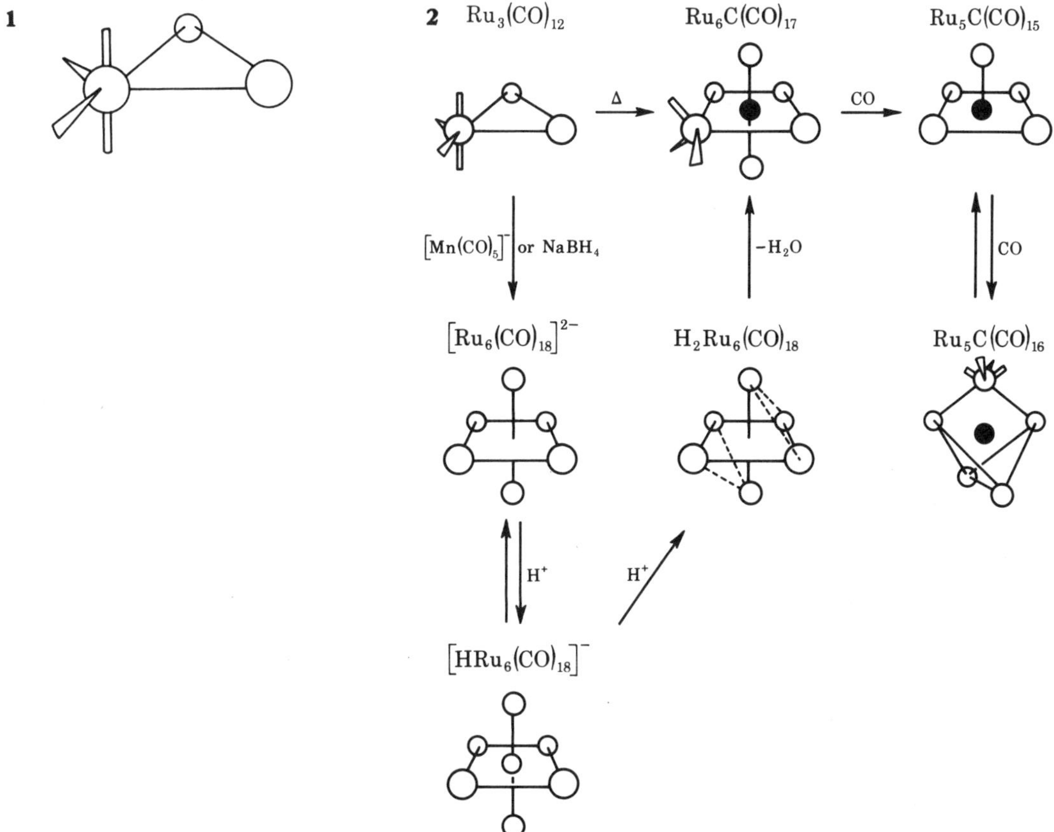

Figure 1. The molecular structure of $Os_3(CO)_{12}$.

Figure 2. Some reactions of $Ru_3(CO)_{12}$ to produce hexaruthenium and pentaruthenium clusters.

INTRODUCTION

In this paper we shall consider the preparation, structure and some chemical properties of cluster carbonyl compounds of both ruthenium and osmium. In figure 1 is illustrated the structure of triosmium dodecacarbonyl, which is the parent of the clusters that we wish to discuss. In this structure, which was originally established by X-ray analysis by Corey & Dahl (1962), the three osmium atoms define an equilateral triangle with three Os–Os distances of *ca.* 288 pm. This distance, which is similar to that of the closely related compound $Ru_3(CO)_{12}$,

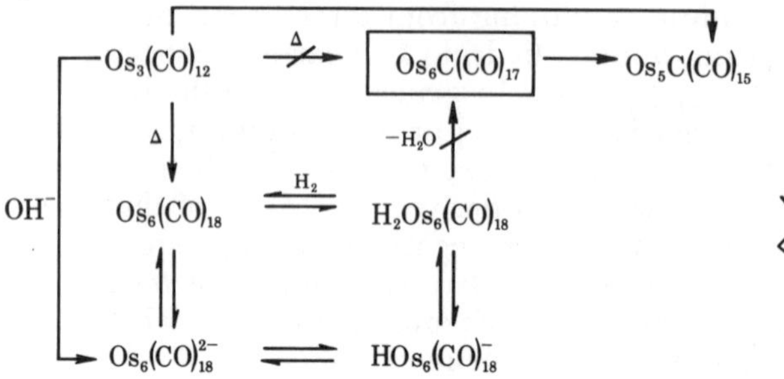

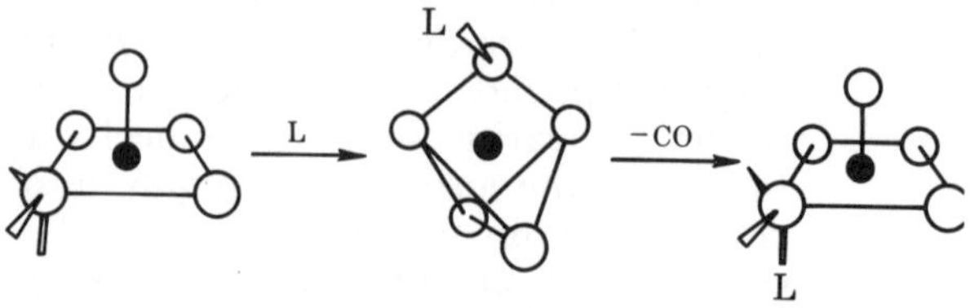

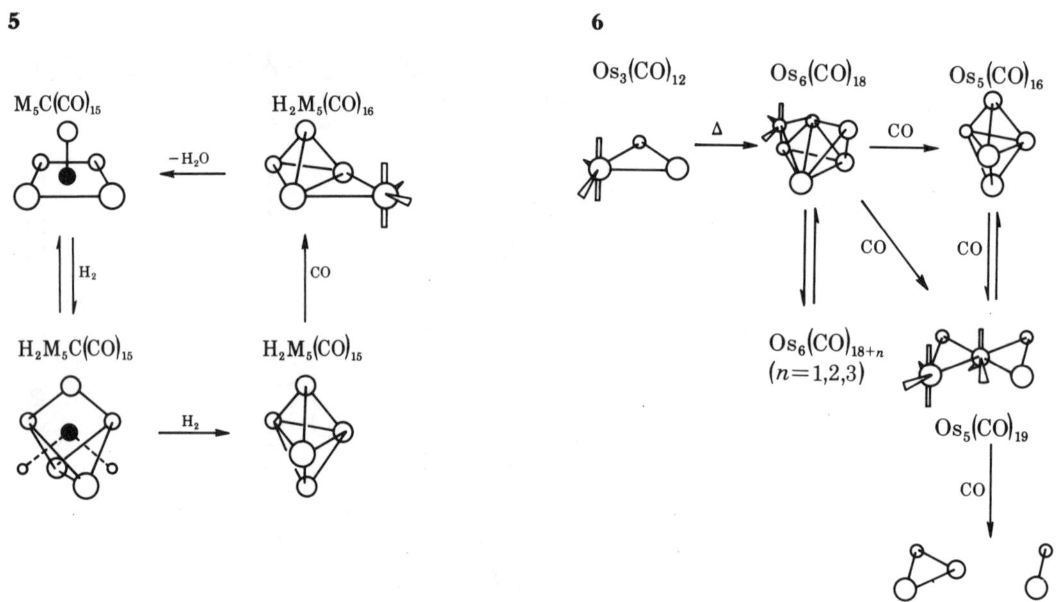

FIGURE 3. Some reactions of $Os_3(CO)_{12}$ to produce hexaosmium and pentaosmium clusters.

FIGURE 4. Possible mechanism of CO substitution in $Ru_5C(CO)_{15}$.

FIGURE 5. The reaction of $Ru_5C(CO)_{15}$ with H_2.

FIGURE 6. Formation and subsequent carbonylation of $Os_6(CO)_{18}$.

is now commonly taken as corresponding to an Os–Os single bond. It is worth noting that within the higher clusters, Os–Os distances ranging from about 260 to 300 pm are observed, and there is no simple relation between bond length and bond multiplicity in systems of this type.

In figure 2 are outlined some of the simple reactions that $Ru_3(CO)_{12}$ will undergo. Several years ago we (Johnson *et al.* 1975) demonstrated that, on pyrolysis, $Ru_3(CO)_{12}$ underwent polymerization and CO_2 ejection to produce $Ru_6C(CO)_{17}$, which has been shown to consist of an octahedron of ruthenium atoms with a central, interstitial carbon atom. More recently we have established that on treatment with CO under moderate pressure and temperature cleavage occurs and the pentanuclear carbide $Ru_5C(CO)_{15}$ and $Ru(CO)_5$ are produced (Farrar *et al.* 1981b). This pentanuclear carbide reacts, in turn, with more CO to form the unstable adduct $Ru_5C(CO)_{16}$ (Johnson *et al.* 1982). This adduct almost certainly possesses a wing-tipped

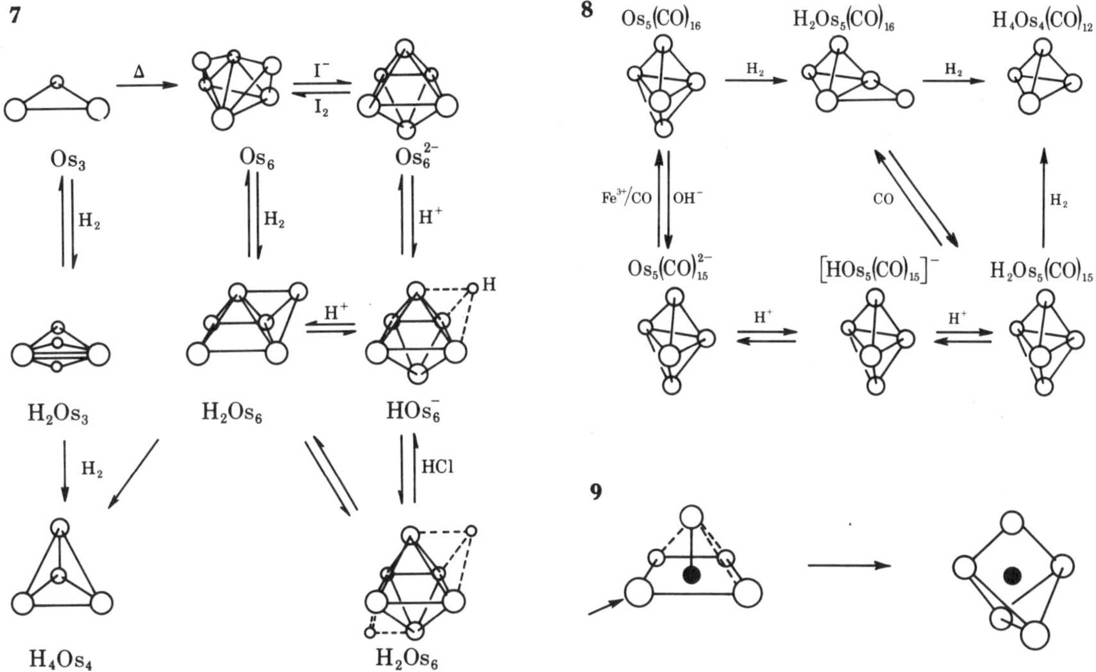

FIGURE 7. The reactions of $Os_3(CO)_{12}$ and $Os_6(CO)_{18}$ with H_2.
FIGURE 8. Formation of pentaosmium and tetraosmium hydrido clusters.
FIGURE 9. Nucleophilic addition to $M_5C(CO)_{15}$.

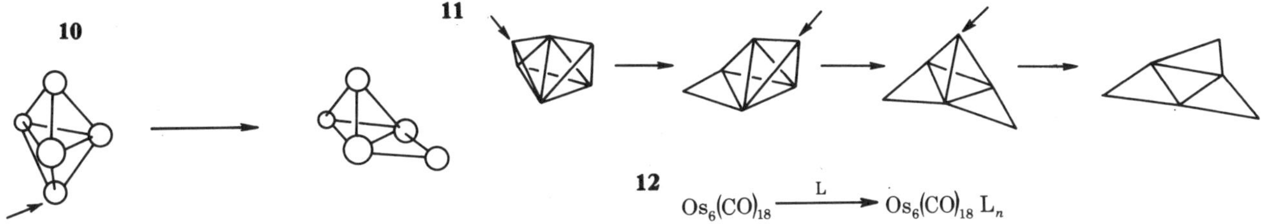

FIGURE 10. Nucleophilic addition to $Os_5(CO)_{16}$.
FIGURE 11. Nucleophilic addition to $Os_6(CO)_{18}$.
FIGURE 12. Addition of ligand L to $Os_6(CO)_{18}$ (L = CO or $P(OMe)_3$; $n = 1$–4).

butterfly structure with an exposed semi-interstitial C atom, by analogy with the structure that we have established for the more stable adduct $Ru_5C(CO)_{15}MeCN$ (Johnson *et al.* 1982), prepared by the direct reaction of $Ru_5C(CO)_{15}$ with MeCN. In an alternative sequence the dianion $Ru_6(CO)_{18}^{2-}$ has been produced from the reaction of $Ru_3(CO)_{12}$ with base; protonation of this dianion produces first $[HRu_6(CO)_{18}]^-$ and then $H_2Ru_6(CO)_{18}$ (Jackson *et al.* 1979a). We find that on heating, water is apparently ejected from this dihydride to provide a second route to $Ru_6C(CO)_{17}$. The corresponding reactions of $Os_3(CO)_{12}$ (figure 3) follow a different course and although varying yields of $Os_5C(CO)_{15}$ (Jackson *et al.* 1980) may be produced from the pyrolysis of $Os_3(CO)_{12}$, $Os_6C(CO)_{17}$ has not been characterized.

The pentanuclear carbides, $M_5C(CO)_{15}$, undergo CO substitution reactions with a range of ligands L to produce $M_5C(CO)_{15-n}L_n$. Almost certainly the mechanism of these reactions is that shown in figure 4. Intermediates of the type $M_5C(CO)_{15}L$ have been fully characterized (see above). Of special note is the reaction of $Ru_5C(CO)_{15}$ with H_2 (figure 5) to produce first $H_2Ru_5C(CO)_{15}$ and then $H_2Ru_5(CO)_{15}$ and (possibly) methane. Further reaction with CO

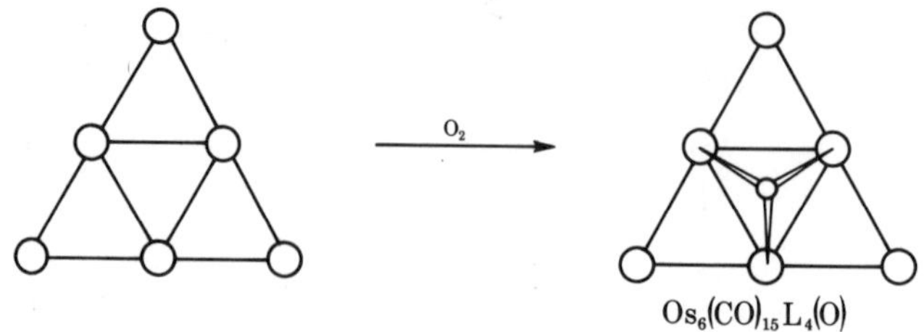

FIGURE 13. Reaction of raft-like clusters $Os_6(CO)_{17}L_4$ with dioxygen.

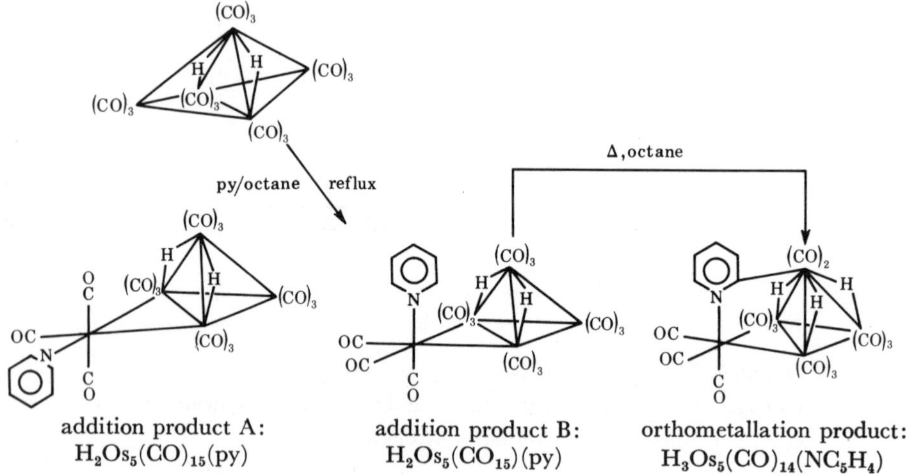

FIGURE 14. Reactions of $H_2Os_5(CO)_{15}$ with pyridine (py).

regenerates $Ru_5C(CO)_{15}$ (Johnson *et al.* 1982). Carbonylation of $Os_6(CO)_{18}$ (figure 6) produces $Os_5(CO)_{16}$ and the new binary carbonyls $Os_5(CO)_{19}$, with a bow-tie metal skeleton, and $Os_6(CO)_{20}$ (Farrar *et al.* 1981*d*). Reaction of H_2 with $Os_6(CO)_{18}$ or $Os_5(CO)_{16}$ generated several hydrido species, but in all cases $H_4Os_4(CO)_{12}$ is eventually produced (figures 7 and 8).

The mechanism of nucleophilic attack on these clusters has not been determined with certainty. However, because the clusters may be regarded as electron-deficient, possible reaction pathways (summarized in figures 9, 10 and 11) would involve addition to the least coordinatively saturated metal atom. For $Os_6(CO)_{18}$ the addition of successive molecules of $P(OMe)_3$ leads to the systematic opening of the bicapped tetrahedral unit, with attack occurring at each unsaturated metal in turn, to produce the planar Os_6 (figures 12 and 13) (Goudsmit *et al.* 1982). The compound $Os_6(CO)_{17}[(P(OMe)_3]_4$ has been shown to react with O_2 to produce $Os_6(CO)_{15}$-$[P(OMe)_3]_4[O]$. Similar nucleophilic additions occur when $H_2Os_5(CO)_{16}$, $Os_5C(CO)_{15}$ or $Os_6(CO)_{18}$ are reacted with pyridine. Here the initial addition is followed by C–H bond cleavage (figures 14, 15 and 16).

Attack of base (OH^-) on $Os_6(CO)_{18}$, $Os_7(CO)_{21}$ or $Os_8(CO)_{23}$ brings about the removal of

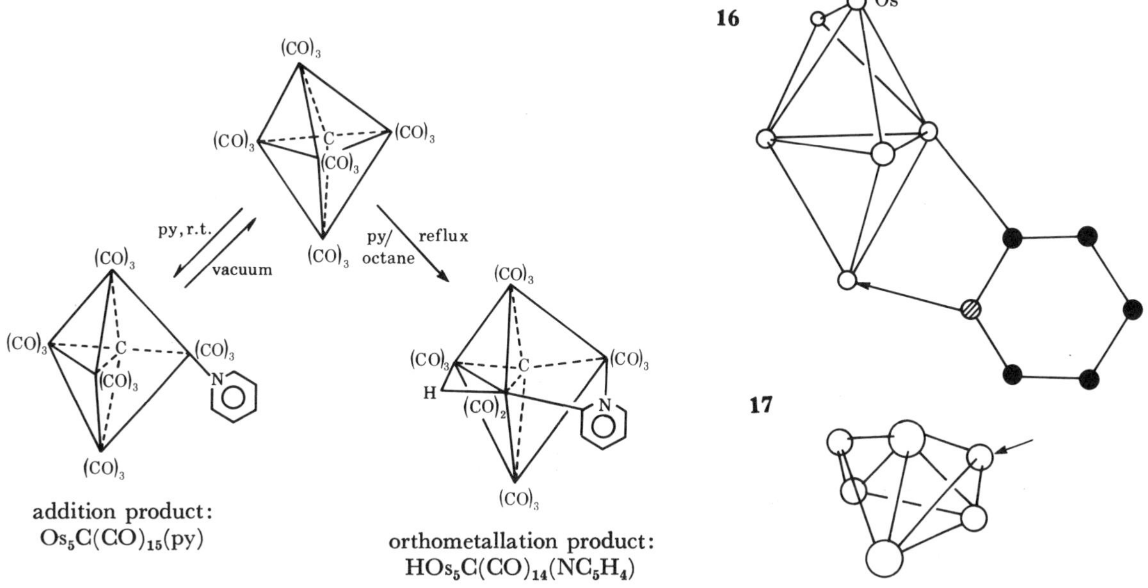

FIGURE 15. Reaction of $Os_5C(CO)_{15}$ with pyridine (py).

FIGURE 16. Arrangement of osmium and ligand anions in $HOs_6(CO)_{16}(C_5H_4N)$.

FIGURE 17. Os_6 core in $Os_6(CO)_{18}$.

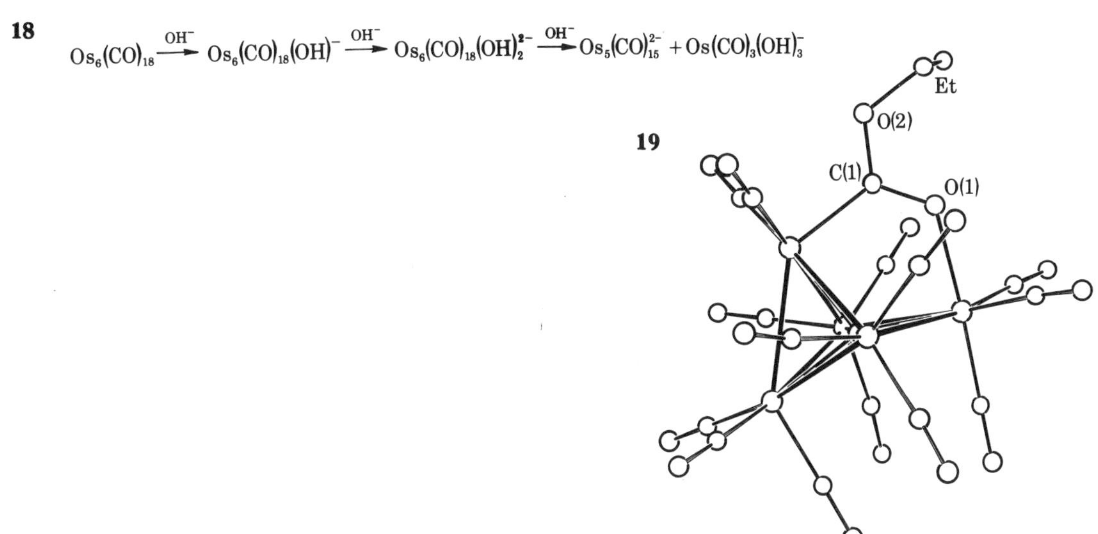

$$18 \qquad Os_6(CO)_{18} \xrightarrow{OH^-} Os_6(CO)_{18}(OH)^- \xrightarrow{OH^-} Os_6(CO)_{18}(OH)_2^{2-} \xrightarrow{OH^-} Os_5(CO)_{15}^{2-} + Os(CO)_3(OH)_3^-$$

FIGURE 18. Addition of OH^- to $Os_6(CO)_{18}$.

FIGURE 19. Molecular structure of $HOs_5C(CO)_{14}(COOEt)$.

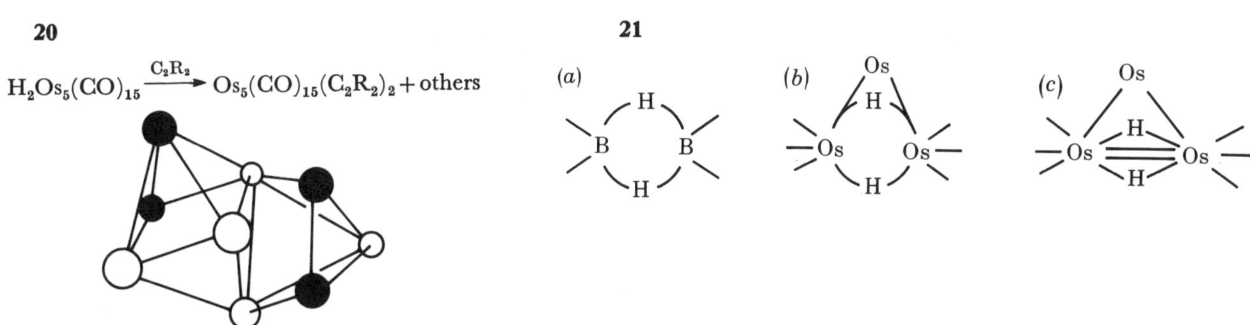

FIGURE 20. Preparation and structure of the organometallic cluster $Os_5(CO)_{15}(C_2R_2)_2$.

FIGURE 21. Bonding picture for $H_2Os_3(CO)_{10}$: (b) analogous to B_2H_6 (a); (c) the alternative viewpoint.

an $Os(CO)_3$ capping unit (figure 17), and here successive nucleophilic addition occurs at the same Os atom (figure 18). $Os_5C(CO)_{15}$ undergoes reaction with ROH in which both nucleophilic addition to a coordinated CO and donation from the so-formed RCO_2 unit to the cluster occurs. The structure of one such derivative (R = Et) is shown in figure 19.

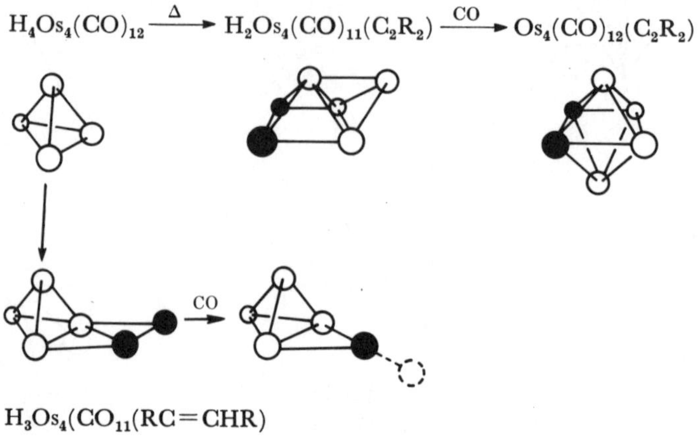

$$H_4Os_4(CO)_{12} \xrightarrow{\Delta} H_2Os_4(CO)_{11}(C_2R_2) \xrightarrow{CO} Os_4(CO)_{12}(C_2R_2)$$

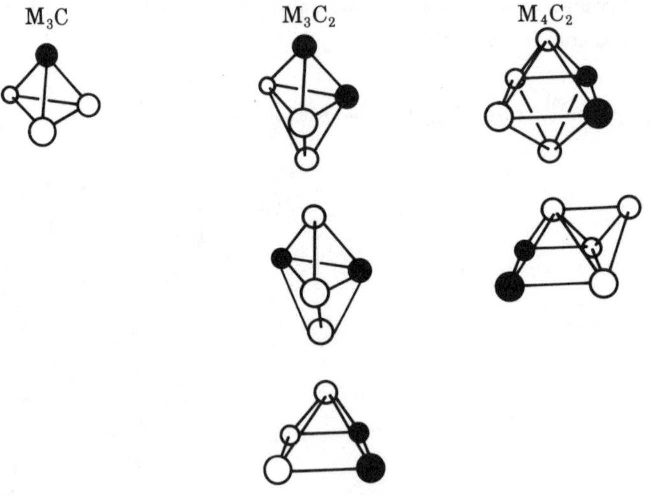

$$H_3Os_4(CO_{11}(RC{=}CHR)$$

FIGURE 22. Reactions of $H_4Os_4(CO)_{12}$ with alkynes (R_2C_2) to produce tetraosmium organometallic clusters.

FIGURE 23. Organo-osmium clusters based on M_3C, M_3C_2 and M_4C_2 units.

tetrahedron $\longrightarrow$ monocapped tetrahedron $\longrightarrow$ bicapped tetrahedron
$H_4Os_4(CO)_{12}$ $[Os_5(CO)_{15}]^{2-}$ $Os_6(CO)_{18}$

FIGURE 24. Tetrahedral growth pattern exhibited by osmium clusters.

O_h octahedron $\longrightarrow$ monocapped octahedron $\longrightarrow$ bicapped octahedron
$[Os_6(CO)_{18}]^{2-}$ $[Os_7(CO)_{21}]^{2-}$ $[Os_8(CO)_{22}]^{2-}$

tetracapped $\longleftarrow$ tricapped
octahedron   octahedron
$[Os_{10}C(CO)_{24}]^{2-}$ '$[Os_9(CO)_{24}]^{2-}$'

FIGURE 25. Growth pattern towards c.c.p. exhibited by osmium clusters.

The similarity between the metal clusters and the boranes is further illustrated by the reactions that they undergo with alkenes and alkynes. Again the reaction pathways are similar to those described above, with initial attack occurring at capping atoms. The structure of $Os_5(CO)_{13}(C_2R_2)_2$ (figure 20) is clearly derived from the trigonal bipyramidal $H_2Os_5(CO)_{15}$ (Farrar *et al.* 1981*c*). This idea of electron-deficiency may be extended to include hydrido-clusters as shown in figure 21, the cluster $H_2Os_3(CO)_{10}$ being more appropriately regarded as having a reaction pattern in keeping with that of B_2H_6. Following this idea that reactions are induced by the opening of three-centre, two-electron Os–H–Os bonds, the behaviour shown by

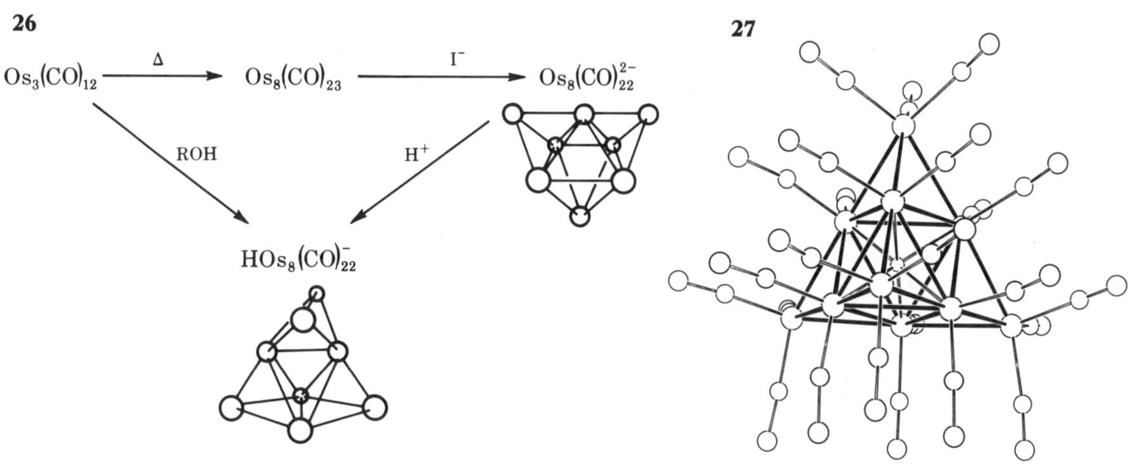

FIGURE 26. Preparation and structure of $[Os_8(CO)_{22}]^{2-}$ and $[HOs_8(CO)_{22}]^-$.

FIGURE 27. Molecular structure of the cluster dianion $[Os_{10}C(CO)_{24}]^{2-}$, emphasizing the close-packed sheath of CO ligands.

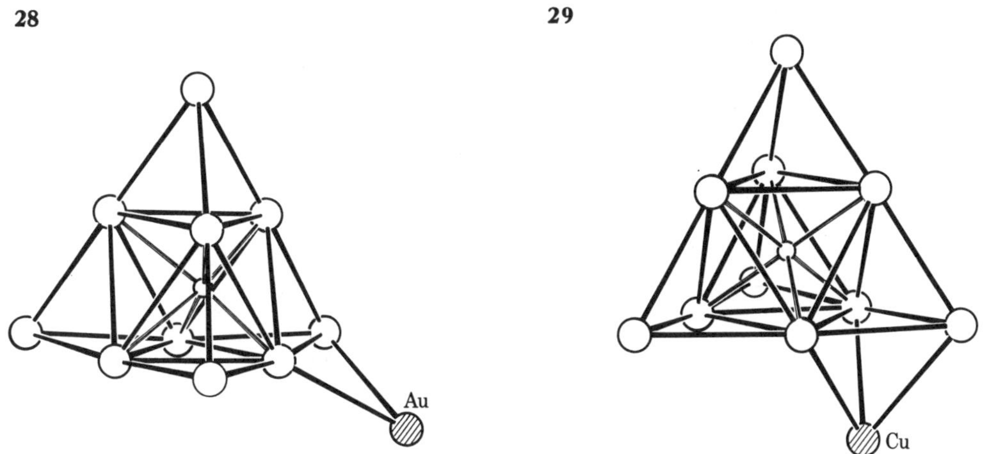

FIGURE 28. Metal geometry observed in the mixed cluster $[Os_{10}C(CO)_{24}Au(PPh_3)]^-$.

FIGURE 29. Metal geometry observed in the mixed cluster $[Os_{10}C(CO)_{24}Cu(NCMe)]^-$.

$H_4Os_4(CO)_{12}$ is readily understood (figure 22) (Johnson *et al.* 1980). Here, octahedral or monocapped square based prismatic geometries are observed in keeping with the view that a CR group is isoelectronic and isolobal with $Os(CO)_3^-$. Other polyhedra based on Os_3C and

Os_3C_2 units may be derived similarly (figure 23). The reactions of $Os_6(CO)_{18}$ are similar, but under the conditions employed expected intermediate compounds have not been observed (Eady *et al.* 1978).

The polyhedral forms adopted by various Os_m fragments fall into two distinct types; those showing the fused tetrahedral growth pattern exemplified by $H_4Os_4(CO)_{12}$, $Os_5(CO)_{15}^{2-}$ and $Os_6(CO)_{18}$ (figure 24), and those exhibiting progressive growth towards c.c.p. (figure 25), $Os_6(CO)_{18}^{2-}$, $Os_7(CO)_{20}^{2-}$, $Os_8(CO)_{22}^{2-}$ and $Os_{10}C(CO)_{24}^{2-}$ are good examples. Other related molecules based on Os_7 and Os_9 units are clearly related to the former. The factors that influence the geometry adopted are not clear. The geometries of the two isoelectronic species

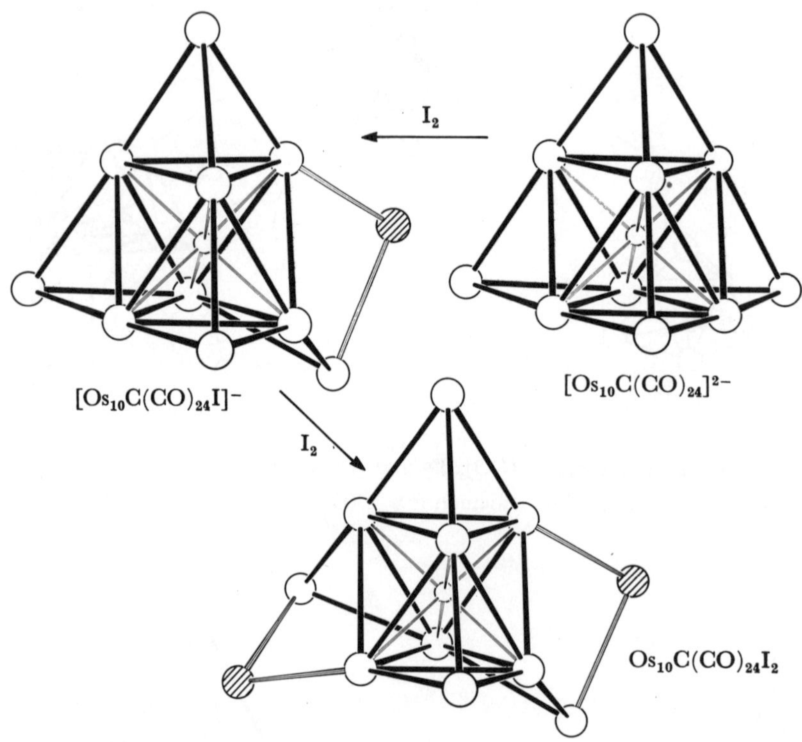

FIGURE 30. Products of the reaction of $[Os_{10}C(CO)_{24}]^{2-}$ with I_2.

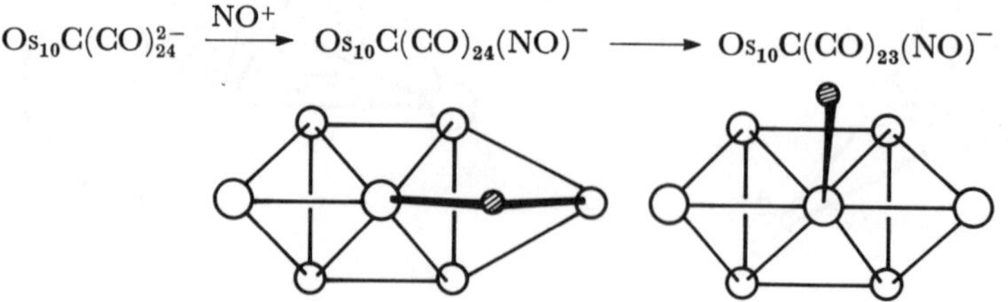

FIGURE 31. Preparation and structure of the nitrosyl cluster compounds
$[Os_{10}C(CO)_{24}(NO)]^-$ and $[Os_{10}C(CO)_{23}(NO)]^-$.

$Os_8(CO)_{22}^{2-}$ and $HOs_8(CO)_{22}^-$ reflect this problem (figure 26). It is possible that the steric bulk of the ligands is important; the inclusion of additional ligands (e.g. H) clearly affects the overall symmetry of the species. Thus the addition of H^+ to the outer ligand sheath of $Os_{10}C(CO)_{24}^{2-}$

(figure 27) would clearly have a marked effect on ligand distribution. In fact an interstitial compound is produced, with the H-ligand occupying a tetrahedral site.

Work now in progress has established that growth along one path or the other may be achieved by the choice of the appropriate bridging or capping atoms. Two examples are shown in figures 28 and 29. These compounds were prepared by the reaction of the appropriate electrophile (e.g. $AuPPh_3^+$) with $Os_{10}C(CO)_{24}^{2-}$. An extensive array of such compounds have now been prepared.

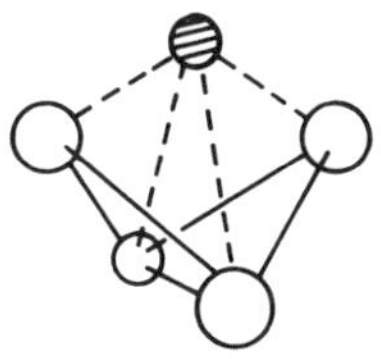

$$HRu_3(CO)_{11}^- \xrightarrow{NO^+} HRu_3(CO)_{10}(NO) \xrightarrow{\Delta} H_nRu_4(CO)_{12}N$$
$$(n = 1, 2 \text{ or } 3)$$

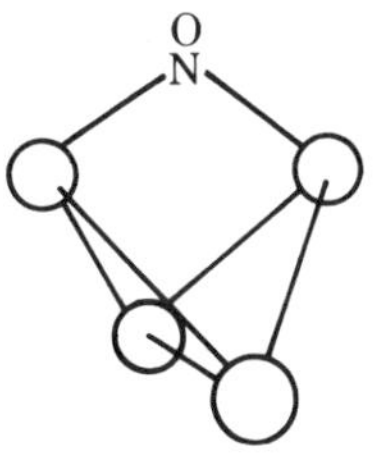

$$H_3Os_4(CO)_{12}^- \xrightarrow{NO^+} H_3Os_4(CO)_{12}(NO) \xrightarrow{\Delta} HOs_4(CO)_{12}N$$

FIGURE 32. Preparation of nitrosyl and nitrido cluster compounds of Ru and Os.

FIGURE 33. The $Os_4(NO)$ arrangement in $H_3Os_4(CO)_{12}NO$.

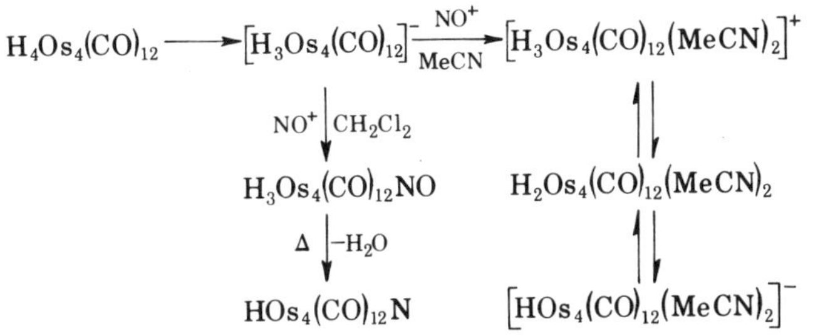

$$H_4Os_4(CO)_{12} \longrightarrow [H_3Os_4(CO)_{12}]^- \xrightarrow[MeCN]{NO^+} [H_3Os_4(CO)_{12}(MeCN)_2]^+$$

$$NO^+ \Big| CH_2Cl_2$$

$$H_3Os_4(CO)_{12}NO \qquad\qquad H_2Os_4(CO)_{12}(MeCN)_2$$

$$\Delta \Big| {-H_2O}$$

$$HOs_4(CO)_{12}N \qquad\qquad [HOs_4(CO)_{12}(MeCN)_2]^-$$

FIGURE 34. Reactions of the anionic cluster $[H_3Os_4(CO)_{12}]^-$ with NO^+ to produce either $H_3Os_4(CO)_{12}NO$ or $[H_3Os_4(CO)_{12}(MeCN)_2]^+$.

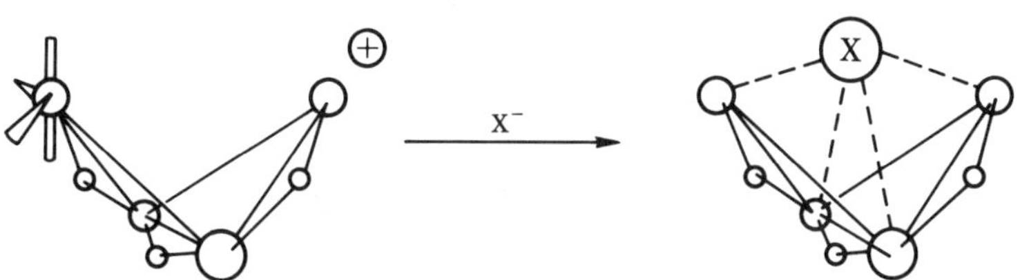

FIGURE 35. Reaction of the cationic cluster $[H_3Os_4(CO)_{12}(MeCN)_2]^+$ with X^- (e.g. Cl^-).

The site of electrophilic attack in cluster species has not been explored in any detail. We have examined some reactions of H^+, I^+, NO^+ and anionic metallic reactants (see above). The reaction of $Os_{10}C(CO)_{24}^{2-}$ with H^+ was discussed earlier, with I^+ the sequence of reactions shown in figure 30 was established (Farrar *et al.* 1981*a*). Very recently we have observed that the same dianion reacts with NO^+ to produce first $Os_{10}C(CO)_{24}(\mu_2\text{-}NO)^-$ and then $Os_{10}C(CO)_{23}(\mu_1\text{-}NO)^-$ (see figure 31). Because the attack of NO^+ directly at a metal atom might be expected to bring about CO loss directly and produce a terminally bound NO species, it is tempting to speculate that attack occurs directly on an Os–Os bond. Other examples of NO[+] addition include the reaction with $H_3Os_4(CO)_{12}^-$ to form either $H_3Os_4(CO)_{12}NO$ (figures 32 and 33) or $[H_3Os_4(CO)_{12}(MeCN)_2]^+$ (figure 34). The former contains a butterfly of four Os atoms with a multibridging NO ligand, and the latter an unsupported butterfly of four osmium atoms. As expected the latter is extremely reactive, combining with a wide variety of reagents X[−] to produce capped butterfly compounds (figure 35).

REFERENCES

Corey, E. R. & Dahl, L. F. 1962 *Inorg. Chem.*, pp. 521–525.

Eady, E. R., Fernandez, J. M., Johnson, B. F. G., Lewis, J., Raithby, P. R. & Sheldrick, G. M. 1978 *J. chem. Soc. chem. Commun.*, pp. 421–423.

Farrar, D. H., Jackson, P. G., Johnson, B. F. G., Lewis, J., Nelson, W. J. H. & Vargas, M. D. 1981*a J. chem. Soc. chem. Commun.*, pp. 1009–1011.

Farrar, D. H., Jackson, P. F., Johnson, B. F. G., Lewis, J. & Nicholls, J. N. 1981*b J. chem. Soc. chem. Commun.*, pp. 415–416.

Farrar, D. H., John, G. R., Johnson, B. F. G., Lewis, J., Raithby, P. R. & Rosales, M. R. 1981*c J. chem. Soc. chem. Commun.*, pp. 886–888.

Farrar, D. H., Johnson, B. F. G., Lewis, J., Nicholls, J. N., Raithby, P. R. & Rosales, M. R. 1981*d J. chem. Soc. chem. Commun.*, pp. 273–274.

Goudsmit, R., Johnson, B. F. G., Lewis, J., Raithby, P. R. & Whitmire, K. 1982 *J. chem. Soc. chem. Commun.* (Submitted.)

Jackson, P. F., Johnson, B. F. G., Lewis, J., McPartlin, M. & Nelson, W. J. H. 1979 *J. chem. Soc. chem. Commun.*, pp. 735–737.

Jackson, P. F., Johnson, B. F. G., Lewis, J., Nicholls, J. N., McPartlin, M. & Nelson, W. J. H. 1980 *J. chem. Soc. chem. Commun.*, pp. 564–565.

Johnson, B. F. G., Kelland, J. W., Lewis, J., Mann, A. L. & Raithby, P. R. 1980 *J. chem. Soc. chem. Commun.*, pp. 547–549.

Johnson, B. F. G., Lewis, J. & Eady, C. R. 1975 *J. chem. Soc. Dalton Trans.*, p. 2606.

Johnson, B. F. G., Lewis, J. & Eady, C. R. J. 1982 (In preparation.)

Discussion

D. M. P. MINGOS (*Inorganic Chemistry Laboratory, University of Oxford, U.K.*). I was most interested to note in Dr Johnson's talk his recent synthesis and structural characterization of an example of a planar 'raft' cluster based on six metal atoms, because independently we have made a theoretical study of the electronic requirements for the stabilization of such cluster compounds. The development of a bonding model for such capped clusters represents an extension of the

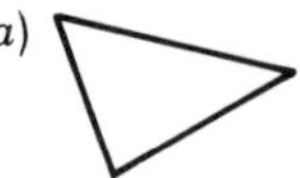 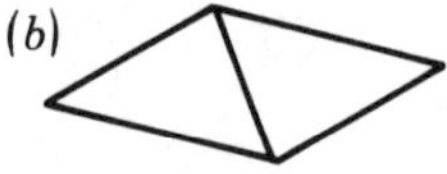 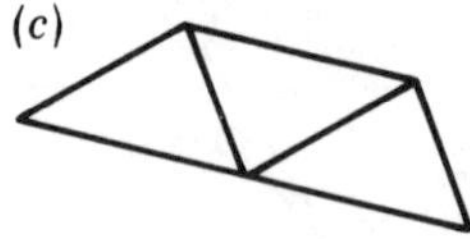 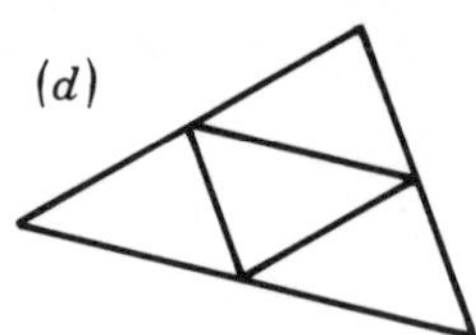

FIGURE D1. Bridged species: (*a*) 48 valence electrons ($Os_3(CO)_{12}$); (*b*) 62 valence electrons ($Re_4(CO)_{16}^{2-}$); (*c*) 76 valence electrons; (*d*) 90 or 92 valence electrons ($Os_6(CO)_{17}(PR_3)_4$).

'capping principle' that we developed several years ago to account for the skeletal electron requirements for capped polyhedra (Mingos 1972; Forsyth & Mingos 1977). The $Os(CO)_4^{2+}$ fragment of C_{2v} symmetry is isolobal with CH_2^{2+} and therefore its bonding capabilities are set by a pair of outpointing hybrid orbitals of a_1 and b_2 symmetry, which are ideally located to enable this fragment to form bonds to a pair of metal atoms and thereby generate a triangular bridge. Therefore from the triangular cluster it is possible to generate the series of bridged species shown in figure D1. The last member in the series has been described as a 'raft cluster' by Dr Johnson. According to the capping principle, the formation of such bridges does not affect the number of bonding molecular orbitals of the parent cluster unless the orbitals of the capping atoms introduce a linear combination of orbitals whose symmetry is not matched by those of the bonding skeletal molecular orbitals of the parent. Since the parent triangular cluster is characterized by a total of 48 valence electrons, it follows that the clusters above will be characterized by a total of $48 + 14m$ valence electrons, where m is the number of bridging $Os(CO)_4$ fragments. The 14 valence electrons per $Os(CO)_4$ fragment are just sufficient to occupy the four $Os–CO$ bonding molecular orbitals and the three non-bonding orbitals of the $Os(CO)_4$ fragment. $Os_3(CO)_{12}$ (48 valence electrons), $Re_4(CO)_{16}^{2+}$ (62 valence electrons) and $Os_6(CO)_{17}(P(OMe)_3)_4$ (90 valence electrons: the compound reported by Dr Johnson) provide examples of (a), (b) and (d) above. This principle can also be applied to open triangular clusters and for example account for the electron count in 'bow-tie clusters'.

It might also be of interest to add that the calculations that we have completed on the raft clusters suggest the presence of a low-lying molecular orbital of a_2' symmetry, which suggests the occurrence of 92-electron clusters in addition to the 90-electron clusters reported. Furthermore, this orbital is antibonding with respect to the metal atoms in the inner triangle and bonding between these metal atoms and the bridging metal atoms, a prediction that could be confirmed by structural studies on related 90-electron and 92-electron clusters.

A detailed account of this work will be presented in a publication by D. G. Evans and myself.

References

Forsyth, M. I. & Mingos, D. M. P. 1977 *J. chem. Soc. Dalton Trans.*, p. 610.
Mingos, D. M. P. 1972 *Nature, phys. Sci.* **236**, 99.

Phil. Trans. R. Soc. Lond. A **308**, 17–26 (1982)   [17]
Printed in Great Britain

Heteronuclear cluster compounds

By H. Vahrenkamp
*Institut für Anorganische Chemie der Universität Freiburg, Albertstrasse 21,
D-7800 Freiburg, F.R.G.*

Several designed syntheses have been developed for heteronuclear organometallic
cluster compounds. They involve step-by-step construction, fragment combination
and metal exchange. The compounds obtained permit the study of the physical
phenomena of electron distribution and chirality in the cluster frameworks. Prominent
chemical phenomena are the weakness of some metal–metal bonds, the unsaturated
nature of some clusters, and polarity due to heteronuclearity. They permit the study
of basic cluster reactions like the unfolding of the clusters with nucleophiles, the
addition of substrates without gross changes in the cluster framework and without
ligand substitution, the capping of clusters by suitable ligands, and the reaction of
different substrates at different locations in the cluster core.

Introduction

Organo-transition metal cluster chemistry has reached a point where systematization has
begun to evolve. This is especially obvious for heteronuclear cluster compounds. Owing to the
availability of suitable reagents and the polarity of hetero-metal–metal bonds their synthesis
can be designed in several ways and their reactivity predicted, at least in part.

Our interest in cluster synthesis and reactivity is one part of our interest in the properties of
metal–metal bonds (Vahrenkamp 1978). Therefore basic metal cluster reactions, as we understand them, involve the formation and rupture of metal–metal bonds, thereby changing the
shape of the cluster core. In most cases we have performed the corresponding chemical reactions on heteronuclear clusters. This paper summarizes our recent results in this area.

Cluster synthesis

The number of specifically designed cluster syntheses, i.e. those where each step in the cluster
build-up is an identifiable product, is still very small. This results from the fact that each
change in the metal atom framework is accompanied by the breaking and reforming of at least
two or three bonds, processes that can even less easily be controlled than those involved in the
construction of carbon-containing cage compounds. In some cases the introduction of multiply
bridging main group elements as constituents of the cluster core improves the situation by
holding the metal atoms together during the critical reactions. We have made use of this clamp
effect in the synthesis and stabilization of compounds with two to four connected metal atoms.
Two simple reactions exemplify this:

$$(CO)_4Fe\!\!-\!\!PH_2\!\!-\!\!Ph \;+\; Co_2(CO)_8 \xrightarrow{\;-CO,\,-H_2\;} (CO)_3Fe\!\!-\!\!P(Ph)\big[Co(CO)_3\big]_2$$

In both cases rather labile starting compounds are converted to rather inert reaction products. The expansion reaction employed in both cases is the metathetical reaction between main-group element hydride functions and metal carbonyl units leading to CO and H_2 evolution and main-group–transition-metal bond formation. This reaction seems to have been first performed by Harrod & Chalk (1965) between Si–H compounds and cobalt carbonyl but has only recently come to the attention of cluster chemists. We could show that it is also applicable for cluster construction from dinuclear starting materials:

$$Fe_2(CO)_6(\mu\text{-PHR})_2$$

$$(R = Me, Ph, t\text{-Bu})$$

And a real stepwise cluster synthesis with the P–H reaction as a crucial step is the following:

$$MePCl_2 \xrightarrow{Me_2NH} MeP(NMe_2)_2 \xrightarrow{Fe_2(CO)_9} (CO)_4Fe\text{–}PMe(NMe_2)_2 \xrightarrow{HCl}$$

By variations of this procedure phosphinidene bridged clusters with one, two, or three different metal atoms can be obtained.

A growing number of mixed metal clusters is being prepared by using clusters as starting materials (Gladfelter & Geoffroy 1980). In most cases the reactions involve cluster expansion. Our contribution in this area is the metal-exchange reaction. It is based on the observation that arsenic-bridged dinuclear complexes (Müller & Vahrenkamp 1977) and simple dinuclear metal carbonyls (Madach & Vahrenkamp 1980) undergo fragmentation and equilibration reactions that can be explained by the intermediate formation of mononuclear metal carbonyl units. Applied to clusters this has allowed the exchange of cobalt for chromium, molybdenum, tungsten, manganese, iron, ruthenium and nickel in an extended range of tetrahedral cluster frameworks. These are composed of tetrametal units or trimetal units with μ_3 bridging CR, SiR, GeR, PR, AsR and S ligands. Among the first systems used for this reaction were the methylidyne tricobalt enneacarbonyls, e.g.:

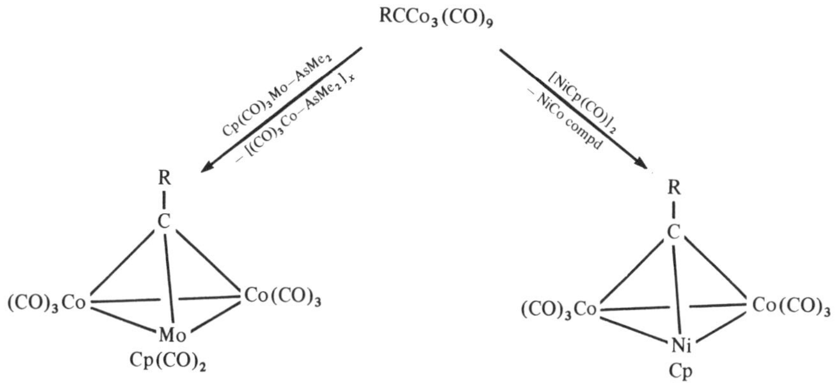

It has turned out that these reactions must be mechanistically quite complex. Carefully controlled conditions are necessary to obtain good yields of the desired products. Otherwise a considerable range of alternative compounds can be formed whose structural features are indicated for two examples in schemes 1 and 2.

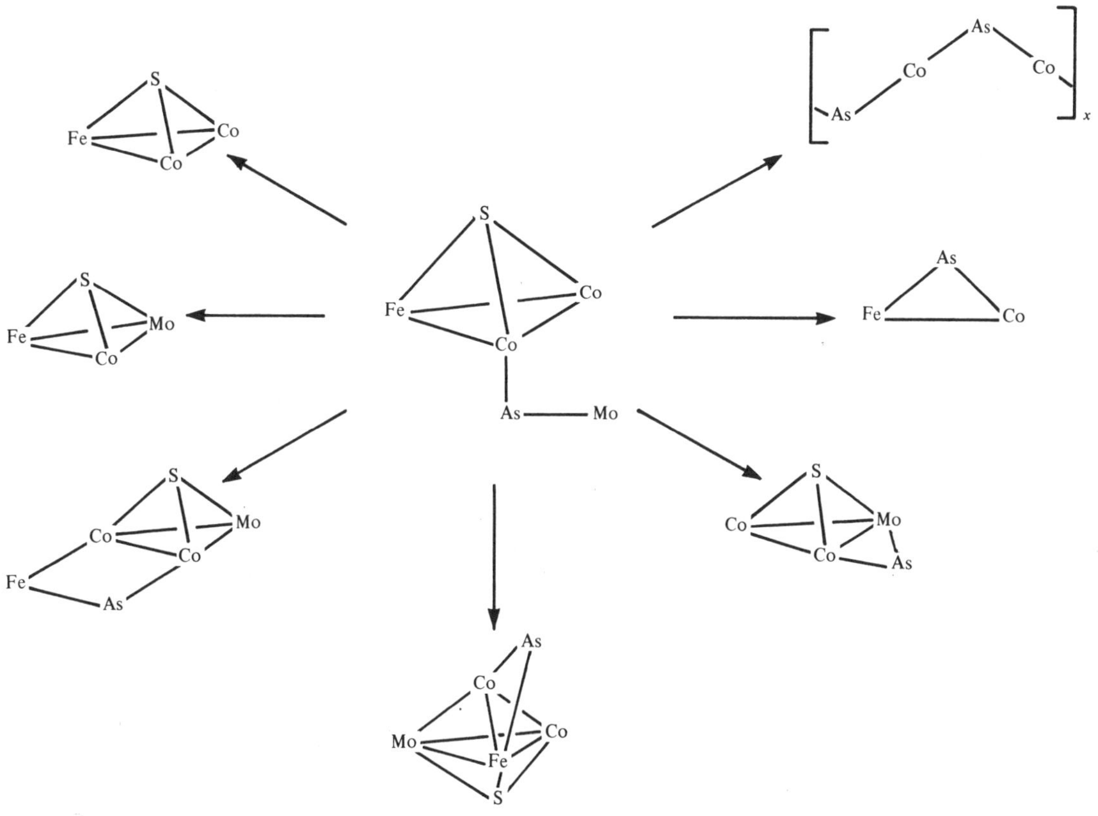

SCHEME 1

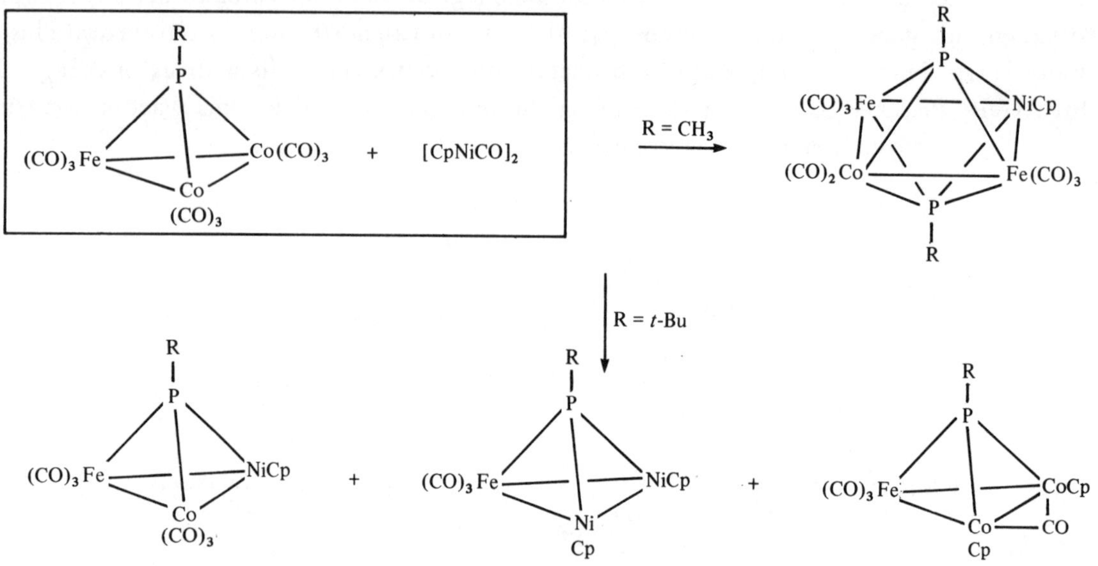

Scheme 2

These alternative reactions support the view that fragmentation is a major characteristic of such first-row transition metal clusters and that the reaction conditions have to limit the fragmentation to the desired amount. However, the large number of new clusters obtained in a predictable way justifies calling this procedure a directed synthesis.

SPECTRAL CLUSTER PROPERTIES RELATED TO THE METAL FRAMEWORK

The accumulation of electrons in the heavy-atom core of transition metal clusters gives rise to electronic phenomena that exceed those of simpler inorganic or organic compounds. Spectral measurements give evidence for this. With the advent of a better theoretical understanding of cluster bonding they should provide a quantitative basis for bonding discussions.

The e.s.r. spectrum of a paramagnetic cluster has been used for a classical qualitative MO description: from an analysis of the 22 line e.s.r. signal of $SCo_3(CO)_9$ Strouse & Dahl (1969) derived the delocalized and metal–metal antibonding nature of the unpaired electron in this cluster. In cooperation with B. H. Robinson, Dunedin, New Zealand, we have extended such e.s.r. measurements to the anions of tetrahedral clusters with one, two and three cobalt atoms. In each case the hyperfine coupling of the electron spin with the cobalt nuclear spin is of the same magnitude indicating comparable electron delocalization, and in each case the fourth framework atom like phosphorus does not show up in the hyperfine structure, indicating the limitation of the h.o.m.o. to the metal region. Figure 1 shows three representative spectra.

Another indication of electron accumulation is given by the o.r.d. spectra of clusters with chiral frameworks. Such clusters were first obtained by us by the aforementioned metal exchange reactions and separated into their enantiomers by substitution with chiral phosphine ligands. As figure 2 shows for SFeCoMo clusters, the o.r.d. spectra of the enantiomers are mirror-symmetric. Those of the diastereoisomers are not, owing to the same configuration of the phosphine in both compounds. The dominance of the metal-containing framework in determining the optical rotation is obvious in several ways: the phosphine part of the diastereo-

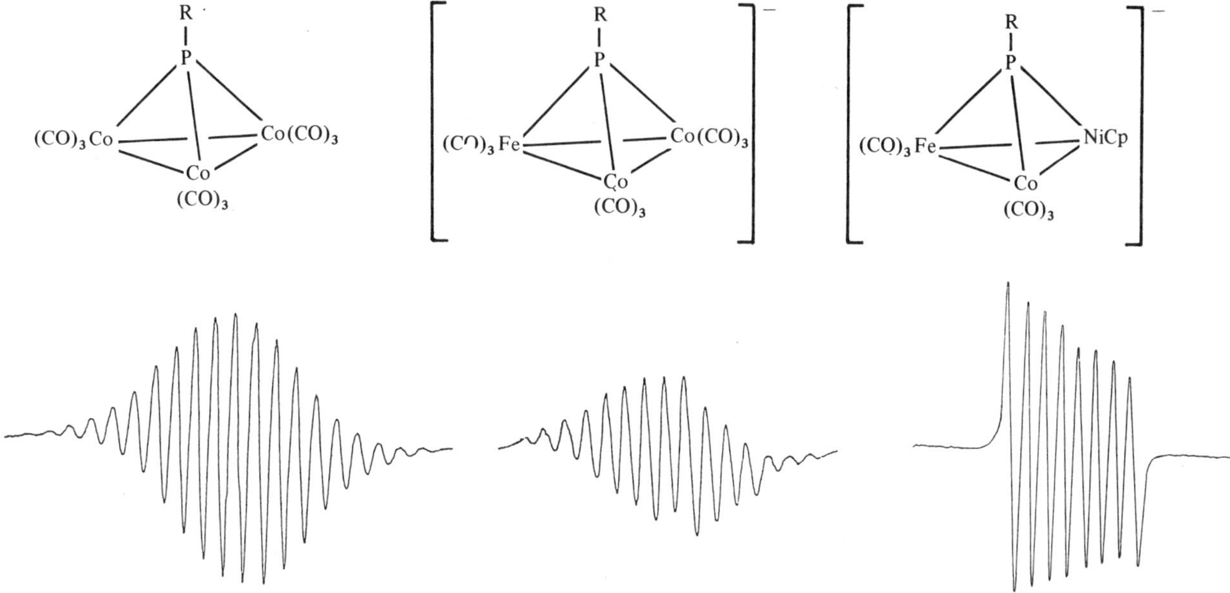

FIGURE 1. E.s.r. spectra of phosphinidene-bridged isoelectronic paramagnetic clusters with three, two, and one cobalt atoms ($I = \frac{7}{2}$), each with one electron in excess of a closed-shell electronic configuration. R = t-Bu, solvent THF, anions prepared electrolytically.

isomers only slightly disturbs the mirror symmetry of their o.r.d. spectra, the appearance of the o.r.d. spectra of the diastereoisomers and of the enantiomers is similar, and moreover the molar rotations are in the order of 10 000–60 000°. Similar cluster types show similar o.r.d. curves, and it is hoped that this can be used to correlate the absolute configurations of such clusters, one of which has been determined.

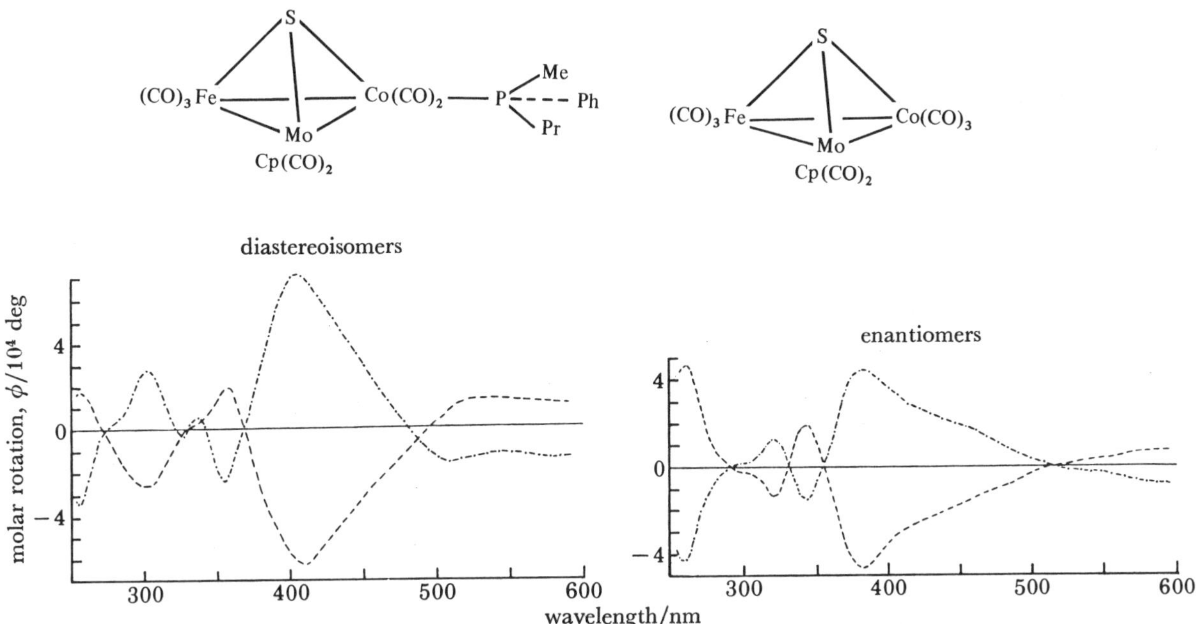

FIGURE 2. Optical rotation of the $(+)$ and $(-)$ isomers of diastereomeric and enantiomeric pairs of SFeCoMo clusters.

In organic chemistry, electron delocalization in ring compounds is indicated by unusual
^{1}H-n.m.r. chemical shifts. Effects of this kind have not yet been observed in metal-containing
rings. We may have found a related phenomenon for the two following cluster types differing
only by one CO ligand (R = Me, t-Bu, p-Tol):

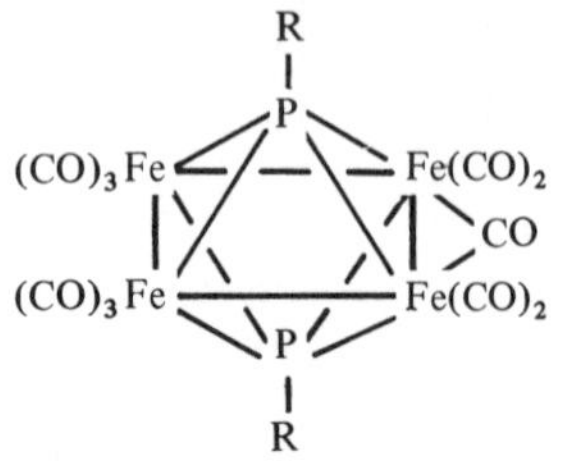

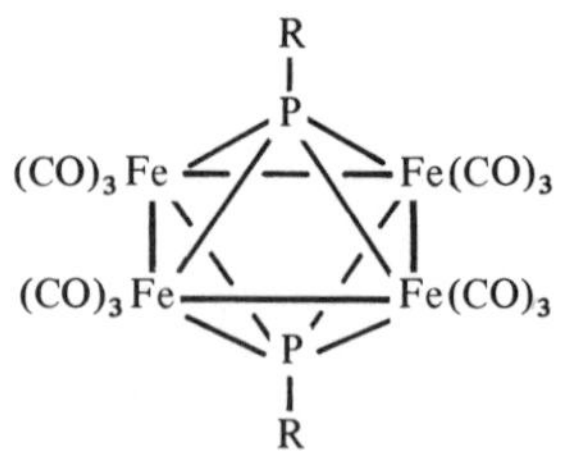

Of these the $Fe_4(CO)_{12}(PR)_2$ compounds are saturated according to the 18-electron rule while
the $Fe_4(CO)_{11}(PR)_2$ compounds lack two electrons. The bond lengths of the latter do not
permit the location of a Fe–Fe double bond unambiguously, which means that they may be
considered as Fe_4 rings with two delocalized π-electrons, i.e. members of the Hückel $4n+2$
series. Support for this comes from the n.m.r. spectra (see table 1). The PR groups located
above the centres of the Fe_4 rings should feel a diamagnetic effect if there is a benzene-type
ring current in the $Fe_4(CO)_{11}(PR)_2$ clusters. There is actually a high field shift for the methyl
resonances in going from the saturated $(CO)_{12}$ clusters to the unsaturated $(CO)_{11}$ clusters. This
shift becomes stronger the closer the observed methyl groups are to the Fe_4 ring.

TABLE 1. ^{1}H-N.M.R. DATA (MILLIONTHS, $CDCl_3$, INT. TMS) FOR THE
METHYL GROUPS OF THE PR LIGANDS IN Fe_4 CLUSTERS

R	$Fe_4(CO)_{12}(PR)_2$	$Fe_4(CO)_{11}(PR)_2$
p-Tol	2.52	2.20
t-Bu	2.06	0.70
Me	3.34	0.98

It may be too early to call these n.m.r. shifts a ring-current phenomenon. But irrespective of
this it cannot be overlooked that the metal atom accumulations do, as might have been predicted, give rise to enhanced n.m.r. effects.

CLUSTER REACTIVITY

Owing to the potential application of cluster compounds in catalysis, many reactions of
organic substrates with metal clusters have been investigated. From an inorganic viewpoint
most of these reactions are variations of the cluster ligand sphere. Comparatively little work
has been performed on the basic cluster reactions, i.e. those that change the geometry or the
bonding in the cluster core. Heteronuclear clusters lend themselves to such reactions owing to
their lower symmetry and their polar metal–metal bonds. We have found examples for several
types of basic cluster reactions.

A very common reaction of first-row transition metal clusters is destruction by strong
nucleophiles causing the rupture of all metal–metal bonds. The reason for this is that for light
transition elements metal–ligand bonding liberates more energy than metal–metal bonding.
For application-oriented considerations it means that the rupture of metal–metal bonds can
provide coordination sites for the activation of substrates. If main-group elements bridge the

cluster metal atoms the identity of the cluster ought not get lost during such an opening of its framework. We (Langenbach & Vahrenkamp 1979) have investigated this in detail for dinuclear arsenic-bridged complexes and found that, even with the simplest nucleophile, CO, reversible opening and closing of the metal–metal bonds are possible. Scheme 3 shows corresponding interconversions for μ_3-PR bridged clusters upon addition or removal of CO. It is obvious that the steric requirements of the metal–ligand units as well as of the PR ligands influence the formation of the preferred compounds. The situation in scheme 4 is somewhat more complicated. Besides demonstrating the reversible unfolding of a tetranuclear cluster it also shows its formation by aggregation and its degradation by Co–As fragment elimination, which all seem to be steps in the above mentioned metal exchange reaction.

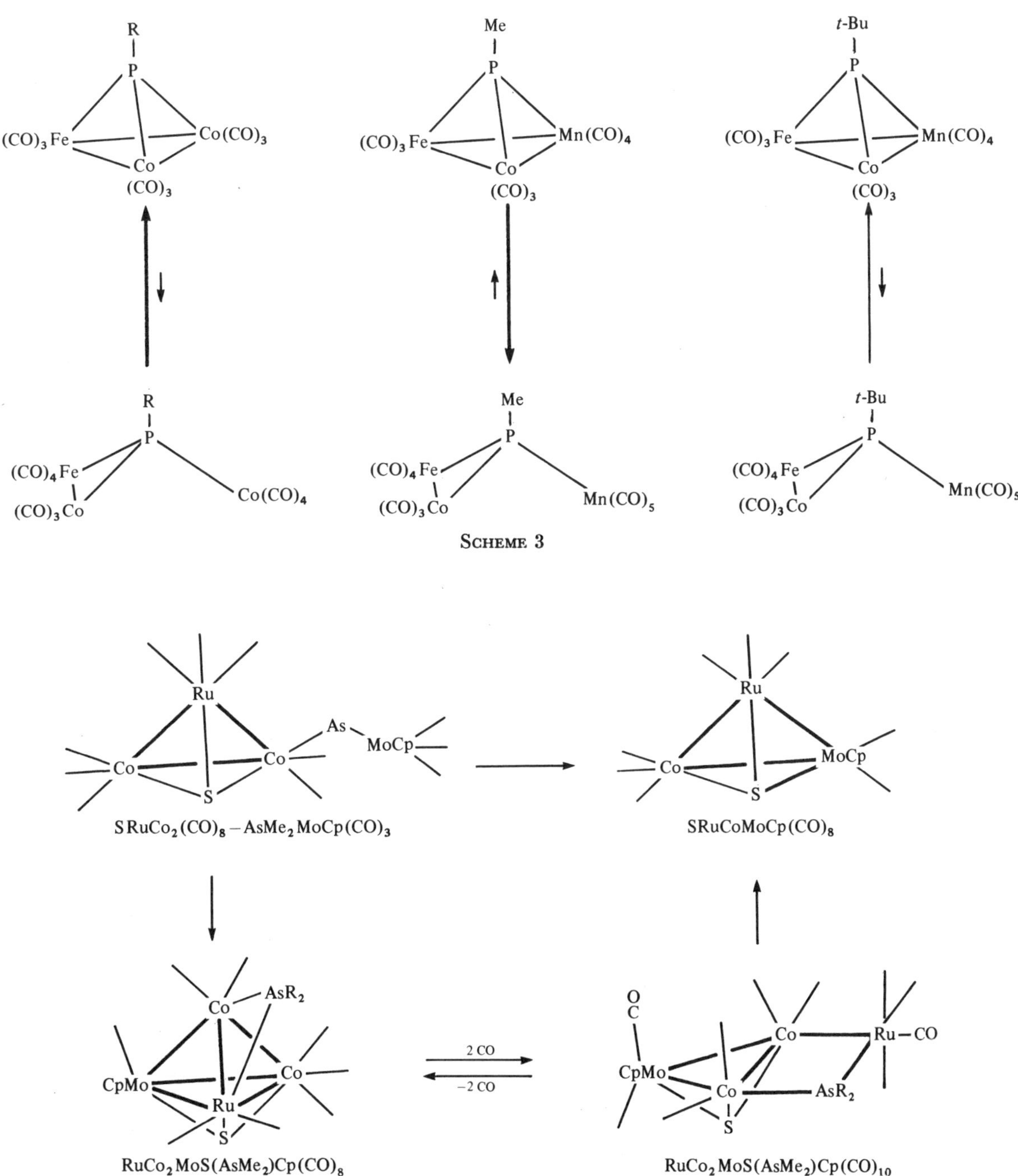

Scheme 3

Scheme 4

Another basic cluster reaction is the addition of nucleophiles to unsaturated clusters without gross changes in the cluster framework and without ligand substitution. We found an entry into the field of unsaturated clusters from the following isoelectronic series of $(\mu_4\text{-PR})_2M_4$ clusters of which the Fe_4 and Fe_2Co_2 compounds have been mentioned above, and of which the Co_4 compound was known before (Ryan *et al.* 1980):

$$(RP)_2Co_4(CO)_{10}, \quad (RP)_2Fe_2Co_2(CO)_{11}, \quad (RP)_2Fe_4(CO)_{12}.$$

In these clusters steric crowding is extensive in the $Fe_4(CO)_{12}$ compound, noticeable in the $Fe_2Co_2(CO)_{11}$ compound and absent in the $Co_4(CO)_{10}$ compound. Relief of the crowding would involve a loss of CO ligands, leading to unsaturation, as in the following clusters:

$$(RP)_2Fe_2Co_2(CO)_{10}, \quad (RP)_2Fe_4(CO)_{11}.$$

Our experiments indicate that the $Fe_2Co_2(CO)_{10}$ compound is very difficult to obtain, whereas the $Fe_4(CO)_{11}$ compound forms easily and quantitatively in vacuum. The unsaturated nature of the latter shows up in the ease of formation of the $Fe_4(CO)_{12}$ compound with CO and the fast addition of other donor ligands. Subsequently another CO ligand can be eliminated and another donor ligand added. For instance, with $L = P(OMe)_3$ the following sequence has been established:

$$(RP)_2Fe_4(CO)_{11} \longrightarrow (RP)_2Fe_4(CO)_{11}L \longrightarrow (RP)_2Fe_4(CO)_{10}L \longrightarrow (RP)_2Fe_4(CO)_{10}L_2.$$

Rather than steric crowding as against unsaturation, electron counting may be invoked to explain the ready interconversions of these Fe_4 clusters: the saturated clusters have the correct electron count according to the 18-electron rule whereas the unsaturated clusters have the correct electron count for *closo* octahedral systems according to Wade's rules.

The two basic cluster reactions described so far provided coordination sites for binding of ligands to one metal atom each. The unique possibilities of clusters, however, rest on their ability to attach one substrate to several metal atoms. Normally clusters have faces of three metal atoms, which may be capped by units with μ_3 bonding ability, i.e. those isolobal with the CH fragment. The basic reaction is then the capping of an M_3 unit to form a tetrahedral framework. We have prepared the cluster $RuCo_2(CO)_{11}$, whose electrophilic nature allows

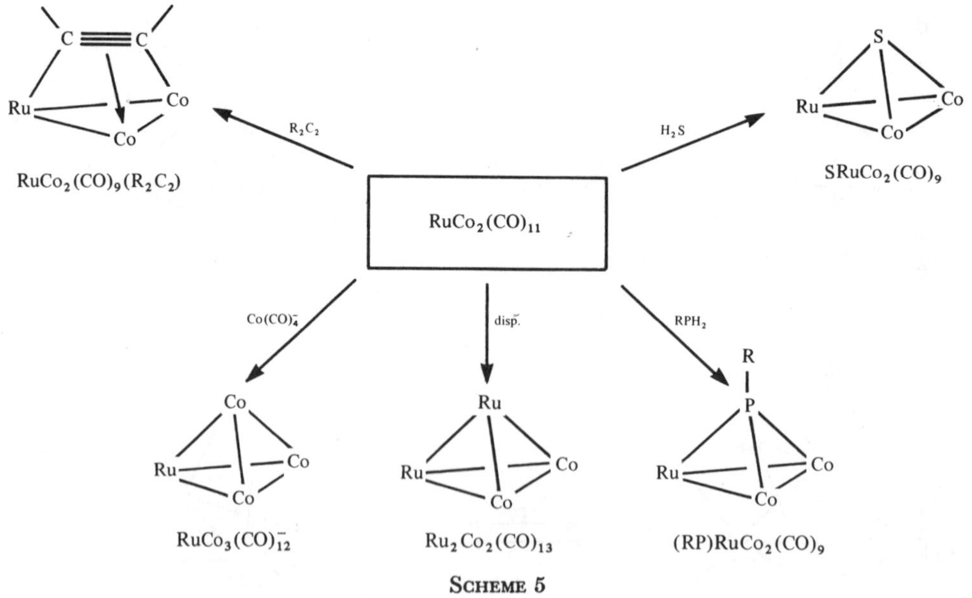

SCHEME 5

capping reactions under very mild conditions. Scheme 5 summarizes the results. The reagents used for capping range from simple four-electron ligands $(RC{\equiv}CR)$ over a series of H_2E compounds $(RPH_2, RAsH_2, SH_2, SeH_2)$ and metal carbonyl anions to simple metal carbonyl fragments like $Ru(CO)_n$ resulting from partial decomposition of the starting cluster. None of these reactions requires heating, and most of them give good yields. Capping undoubtedly is an elemental step in cluster build-up reactions, but the number of cases where simple capping products could be isolated is still relatively small.

Finally, multimetal reactivity should be considered. One of the advantages of heteronuclear clusters is their ability to activate different substrates at different metal atoms. This gives rise to the hope of the specific catalysis of complicated organic reactions. Again the number of examples to underline this is quite small. During our studies of ruthenium–cobalt clusters we found a partial approach to a result of this type. The two reactions are:

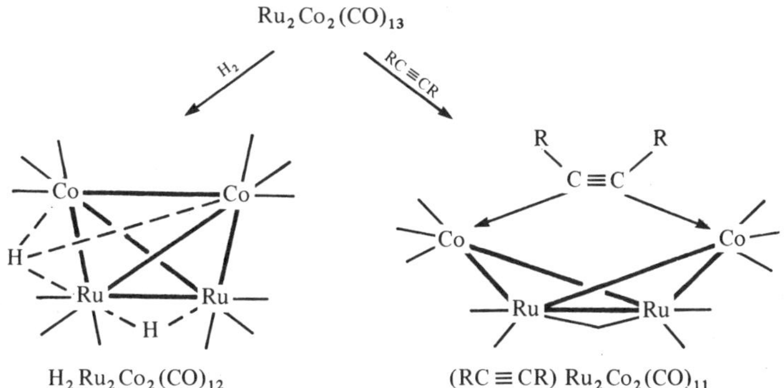

The affinity of cobalt carbonyl compounds for acetylenes has long been known, and butterfly-shaped $(RC{\equiv}CR)$ $Co_4(CO)_{10}$ was prepared more than 20 years ago (Hübel *et al.* 1959). Conversely, ruthenium carbonyls react easily with hydrogen (Knox *et al.* 1975). Both properties are combined in $Ru_2Co_2(CO)_{13}$: acetylenes insert between the cobalt atoms, and hydrogen is added at the ruthenium atoms. Both reactions require mild conditions and produce essentially one compound. The continuation of the idea, however, is not simple: either the acetylene cluster and hydrogen or the hydrogen cluster and acetylenes react to produce mixtures that have not been separated so far.

The basic cluster reactions here, which also include the metal-exchange reaction, are just a beginning. The immense number of clusters synthesized so far offers many more possibilities, which have yet to be realized.

This work was supported by the Deutsche Forschungsgemeinschaft and by the Fonds der Chemischen Industrie. It was performed by a team of hard-working graduate students: E. Keller, H. J. Langenbach, E. Röttinger, T. Madach, H. Beurich, F. Richter, U. Honrath, M. Müller, E. Roland, K. Fischer, P. Gusbeth and R. Blumhofer. The cooperation with Professor B. H. Robinson, Dunedin, New Zealand, and his research group is gratefully acknowledged.

REFERENCES

Gladfelter, W. L. & Geoffroy, G. L. 1980 *Adv. organometall. Chem.* **18**, 207–274.
Harrod, J. F. & Chalk, A. J. 1965 *J. Am. chem. Soc.* **87**, 1133–1135.
Hübel, W., Braye, E. H., Clauss, A., Weiss, E., Krüerke, U., Brown, D. A., Kings, G. S. D. & Hoogzand, C. 1959 *J. inorg. nucl. Chem.* **9**, 204–210.

Knox, S. A. R., Koepke, J. W., Andrews, M. A. & Kaesz, H. D. 1975 *J. Am. chem. Soc.* **97**, 3942–3947.
Langenbach, H. J. & Vahrenkamp, H. 1979 *Chem. Ber.* **112**, 3773–3794.
Madach, T. & Vahrenkamp, H. 1980 *Chem. Ber.* **113**, 2675–2685.
Müller, R. & Vahrenkamp, H. 1977 *Chem. Ber.* **110**, 3910–3919.
Ryan, R. C., Pittman, C. U., O'Connor, J. P. & Dahl, L. F. 1980 *J. organometall. Chem.* **193**, 247–269.
Strouse, C. E. & Dahl, L. F. 1969 *Discuss. Faraday Soc.* **47**, 93–106.
Vahrenkamp, H. 1978 *Angew. Chem.* **90**, 403–416; *Angew. Chem. int. Edn Engl.* **17**, 379–392.

Phil. Trans. R. Soc. Lond. A **308**, 27–34 (1982)   [27]

Printed in Great Briatin

The chemistry of stabilized clusters

By J. A. Iggo, M. J. Mays and P. L. Taylor

*University Chemical Laboratory, Lensfield Road,
Cambridge, CB2 1EW, U.K.*

The complexes $[Ru_3(\mu_2\text{-}H)_2(\mu_3\text{-}PPh)(CO)_9]$ (**1**) and $[Mn_2(\mu_2\text{-}H)(\mu_2\text{-}PPh_2)(CO)_8]$ (**2**) have been prepared in high yields from, respectively, the reaction of $[Ru_3(CO)_{12}]$ with $PhPH_2$ and the reaction of $[Mn_2(CO)_{10}]$ with Ph_2PH. Complexes **1** and **2** undergo a range of reactions without fragmentation into mononuclear species. Thus **1**, after treatment with Me_3NO, reacts with a wide range of two-electron donor ligands, L, to give the substitution products $[Ru_3(\mu_2\text{-}H)_2(\mu_3\text{-}PPh)(CO)_8L]$. It also undergoes oxidative addition reactions with addenda XY to give the complexes $[Ru_3(\mu_2\text{-}H)_2(X)(Y)(\mu_3\text{-}PPh)(CO)_9]$ (**3**), $[Ru_3(\mu_2\text{-}H)_2(Y)(\mu_2\text{-}X)(\mu_3\text{-}PPh)(CO)_8]$ (**4**) and $[Ru_3(\mu_2\text{-}H)_2(\mu_2\text{-}Y)(\mu_2\text{-}X)(\mu_3\text{-}PPh)(CO)_7]$ (**5**) (**3**, X = Y = Cl or Br; **4**, X = Y = Cl or Br; X = Cl, Y = HgCl; **5** *a*, X = Y = Cl or Br; **5** *b*, X = Cl, Y = HgCl). $[Mn_2(\mu_2\text{-}H)(\mu_2\text{-}PPh_2)(CO)_8]$ (**2**) undergoes substitution reactions to give complexes of formula $[Mn_2(\mu_2\text{-}H)(\mu_2\text{-}PPh_2)(CO)_{8-n}L_n]$ (n = 1, 2 or 4) and reacts with alkynes, $RC\!:\!CR'$, to give the complexes $[Mn_2(\mu_2\text{-}\eta^2\text{-}RC\!:\!C(H)R')(\mu_2\text{-}PPh_2)(CO)_7]$, which themselves react further with electron donor molecules such as R_3P and H^- to give μ_2-carbene complexes. Complexes **1** and **2** react with Na/Hg to give, respectively, the anions $[Ru_3(\mu_2\text{-}H)(\mu_3\text{-}PPh)(CO)_9]^-$ and $[Mn_2(\mu_2\text{-}PPh_2)(CO)_8]^-$, which react with neutral or cationic complexes of other metals to give a variety of mixed-metal clusters.

Introduction

Studies of the chemistry of metal cluster complexes and, in particular, their reactions with small organic molecules, have been confined to relatively few systems. Among the reasons for this are:

(i) not many clusters are easily synthesized in high yields;

(ii) their reactions often give a multitude of products that are difficult to separate and characterize;

(iii) the conditions required to bring about reactions often lead to fragmentation of the cluster into lower nuclearity (often mononuclear) species.

One cluster whose chemistry has been extensively studied is $[Os_3H_2(CO)_{10}]$. This can be synthesized in high yields from $[Os_3(CO)_{12}] + H_2$ (Knox *et al.* 1975) and reacts readily under mild conditions with a wide range of electron-donor molecules by virtue of its coordinative unsaturation (Shapley *et al.* 1975; Deeming & Hasso 1976; Adams & Golembeski 1979). Formally, one may consider that a metal–metal double bond is present, which is reduced to a single bond on coordination of an additional two-electron donor ligand such as an organophosphine. The presence of metal–hydrogen bonds in this cluster and the cluster's ability to coordinate organic substrates enable it to undergo a wide variety of insertion reactions, leading to products that may be regarded as intermediates in the reduction of organic molecules by clusters (Deeming & Hasso 1975; Keister & Shapley 1975).

Results and discussion

Our interest in this area began with a study of the interaction of $[Os_3H_2(CO)_{10}]$ with various fluorocarbon molecules in an attempt to delineate cyclic reduction processes by using clusters. In some cases, such as the hydrogenation of an alkyne to an alkene, cyclic processes can be completed (Dawoodi *et al.* 1982). In other examples, such as the attempted hydrogenation of a nitrile, the stability of an intermediate complex prevents completion of the cycle (Banford *et al.* 1982).

There is a small number of polynuclear transition metal complexes related to $[Os_3H_2(CO)_{10}]$ in that metal–hydrogen and multiple metal–metal bonds (in the same sense as in $[Os_3H_2(CO)_{10}]$) are present. In general, however, these complexes are more difficult to synthesize in high yield than $[Os_3H_2(CO)_{10}]$ and, more significantly, they decompose readily into lower nuclearity fragments on treatment with electron-donor ligands (Andrews *et al.* 1977):

$$(OC)_4Re \overset{H}{\underset{H}{=\!=\!=}} Re(CO)_4 + 2Ph_3P \rightarrow 2[HRe(CO)_4(PPh_3)].$$

We have sought to prepare complexes of this type containing bridging ligands and, for example, in the complexes $[Re_2(\mu_2\text{-}H)_2(CO)_6(LL)]$ (LL = $Ph_2PCH_2PPh_2$ or $(EtO)_2POP\text{-}(OEt)_2$), we have found that such ligands do stabilize the dinuclear unit with respect to the above decomposition (Mays *et al.* 1980). With suitable organic substrates, insertion reactions, analogous to those of $[Os_3H_2(CO)_{10}]$, can then be observed (figure 1).

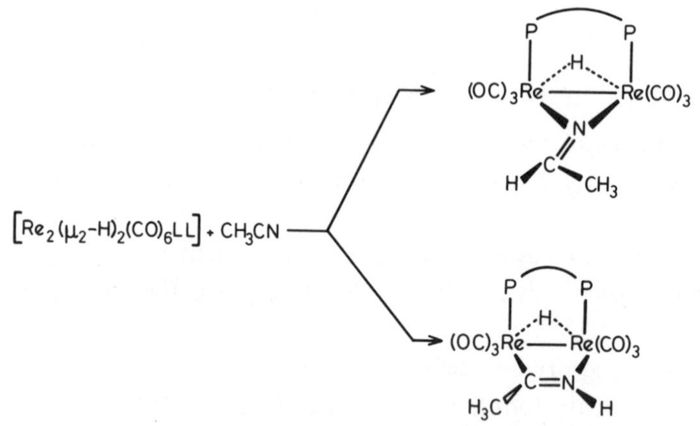

FIGURE 1. The reaction of $[Re_2(\mu_2\text{-}H)_2(CO)_6(LL)]$ with CH_3CN (LL = $Ph_2PCH_2PPh_2$ or $(EtO)_2POP(OEt)_2$).

The above examples of insertion reactions involve coordinatively unsaturated polynuclear complexes, and in the course of the reaction a formal hydrogen-bridged metal–metal bond of order two is reduced to a single bond, which presumably facilitates the initial coordination of the organic substrate. Recent work by, for example, Huttner *et al.* (1979) shows, however, that the coordination of electron-donor ligands to clusters can also be achieved by a reduction of metal–metal bond order from one to zero, in particular for clusters containing first-row transition metals. This suggests that apparently coordinatively saturated 'stabilized' clusters containing hydrogen-bridged metal–metal bonds might react with electron donor ligands in a similar way (figure 2). To examine this possibility we have investigated the reactions of some complexes of this type, concentrating on those for which we have been able to devise high-yield syntheses. One such complex is $[Ru_3(\mu_2\text{-H})_2(\mu_3\text{-PPh})(CO)_9]$ (1), which may be obtained in 50% yield from the reaction of $[Ru_3(CO)_{12}]$ with $PhPH_2$ in refluxing hexane (Iwasaki *et al.* 1981). Treatment of this complex with electron donor ligands, however, does not result in fission of the metal–metal or hydrogen-bridged metal–metal bonds. Thus, under a pressure of 50 atm (1 atm $\approx$ 133 Pa) CO at 100 °C in heptane solution, no change in the infrared spectrum of the cluster is observed. With phosphorus donor ligands (L) the substitution products, $[Ru_3(\mu_2\text{-H})_2(\mu_3\text{-PPh})(CO)_8L]$ rather than simple adducts, are obtained on refluxing in cyclohexane. A much wider range of substitution products (table 1) can be obtained in high yield by treatment of 1 with Me_3NO in CH_2Cl_2 at room temperature followed by addition of the ligand.

FIGURE 2. Suggested mode of reaction of complexes containing hydrogen-bridged metal–metal bonds with electron donor ligands, L′ (X is a bridging ligand).

TABLE 1. SUBSTITUTION PRODUCTS DERIVED FROM $[Ru_3(\mu_2\text{-H})_2(\mu_3\text{-PPh})(CO)_9]$

compound†	$\nu(CO)/cm^{-1}$‡	^1H n.m.r.§
$[Ru_3(\mu_2\text{-H})_2(\mu_3\text{-PPh})(CO)_8(PPh_3)]$	2073m, 2043s, 2036s, 2003s, 1989s, 1980m, 1967w	7.53(m, 20H, Ph), -18.54(d of d, $^2J_{PH} = 16$ Hz, $^2J_{P'H} = 10$ Hz, 2H, RuH)
$[Ru_3(\mu_2\text{-H})_2(\mu_3\text{-PPh})(CO)_8(PH_2Ph)]$	2077m, 2048s, 2041s, 2009s, 2001m, 1993m, 1978w	7.70(m, 10H, Ph), 5.97 (d, $J_{PH} = 354$ Hz, 2H, Ph), -19.02 (d of d, $^2J_{PH} = 14$ Hz, $^2J_{P'H} = 15$ Hz, 2H, RuH)
$[Ru_3(\mu_2\text{-H})_2(\mu_3\text{-PPh})(CO)_8(CNBu^t)]$	2072m, 2046s, 2036s, 2004s, 1995m, 1981m, 1969w	7.50(m, 5H, Ph), 1.33(s, 9H, Bu^t) -18.88(d, $^2J_{PH} = 16$ Hz, 2H, RuH)
$[Ru_3(\mu_2\text{-H})_2(\mu_3\text{-PPh})(CO)_8(NCMe)]$	2076m, 2043s, 2035s, 2002s, 1992m, 1979m, 1967w	
$[Ru_3(\mu_2\text{-H})_2(\mu_3\text{-PPh})(CO)_8(\mu_2\text{-CH}_2)]$	2087m, 2056s, 2035s, 2010m, 1987m, 1968w, 1767w, br	7.55(m, 5H, Ph), 5.96(m, 1H, CH_2), 4.18(m, 1H, CH_2), -17.45(br d, $^2J_{PH} = 15$ Hz, 2H, RuH)

† Molecular ion peaks observed in mass spectra of all compounds except $[Ru_3(\mu_2\text{-H})_2(\mu_3\text{-PPh})(CO)_8(NCMe)]$.
‡ Recorded in cyclohexane solution.
§ In millionths relative to $SiMe_4$ and recorded in CD_2Cl_2 solution at -70°C.

An alternative approach to bond breaking is to treat the complex with reducing agents. Reaction of 1 with sodium amalgam at room temperature, however, gives the anion $[Ru_3H(CO)_9PPh]^-$ in quantitative yield:

$$[Ru_3(\mu_2\text{-H})_2(\mu_3\text{-PPh})(CO)_9] \xrightarrow{\text{Na/Hg}} [Ru_3(\mu_2\text{-H})(\mu_3\text{-PPh})(CO)_9]^- + \tfrac{1}{2}H_2.$$

Oxidative addition can also lead to bond fission in polynuclear complexes and treatment of 1 with suitable addenda does lead to initial cleavage of the lone metal–metal bond, although the hydrogen-bridged metal–metal bonds remain intact. Thus if a solution of 1 in CH_2Cl_2 at room temperature is allowed to react with Cl_2 or Br_2 a mixture of complexes (3, 4 and 5*a*) is obtained after a few seconds (figure 3). With $HgCl_2$ the initial products are presumably analogous to **3** and **4**, but the major product observed even at room temperature is **5b**, in which the ruthenium–ruthenium bond has reformed. A product analogous to **4** has been isolated by Adams (1982) from the reaction of $[Ru_3(\mu_2\text{-}H)_2(CO)_9S]$ with $SnCl_4$, and type **5b** complexes have been obtained previously from the reaction of $[Ru_3H(CO)_9C_6H_9]$ with PhHgX (X = Br, I) (Fahmy *et al.* 1980).

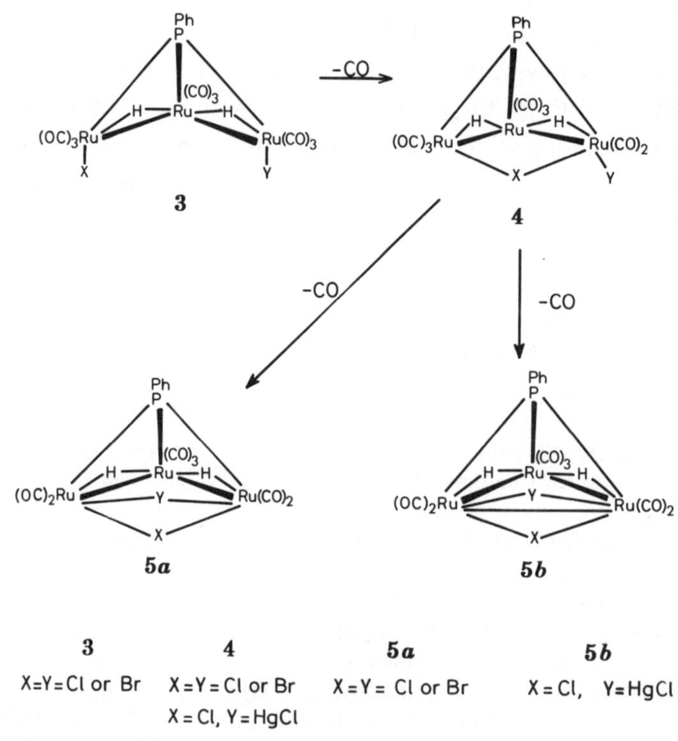

3	4	5a	5b
X = Y = Cl or Br	X = Y = Cl or Br	X = Y = Cl or Br	X = Cl, Y = HgCl
	X = Cl, Y = HgCl		

FIGURE 3. Proposed structures for oxidative addition products from reaction of $[Ru_3(\mu_2\text{-}H)_2(\mu_3\text{-}PPh)(CO)_9]$ with addenda XY.

Previous studies indicate that metal–metal bonds involving first row transition metals are more easily broken than those involving second and third row transition elements, and it seemed likely that the same would be true of hydrogen-bridged metal–metal bonds. We have therefore also studied the reactivity of the complex $[Mn_2(\mu_2\text{-}H)(\mu_2\text{-}PPh_2)(CO)_8]$ (**2**). This was first prepared by Green & Moelwyn-Hughes (1962) in 2% yield as a by-product from the reaction of Ph_2PCl with $[Mn(CO)_5]^-$ and subsequently by Hayter (1964) in 12% yield from the reaction of $[Mn_2(CO)_{10}]$ with Ph_4P_2. We have now synthesized **2** in 80% yield by treatment of $[Mn_2(CO)_{10}]$ with Ph_2PH in undried decalin at 150 °C. As with **1**, however, high-pressure infrared studies in heptane solution reveal that **2** is unaffected by CO even at pressures of 50 atm and temperatures of 200 °C. Reaction of **2** with other two-electron donor ligands such as organophosphines, nitriles or isonitriles in refluxing cyclohexane or on u.v. irradiation in pentane solution results in the formation of substitution products of general formulae $[Mn_2(\mu_2\text{-}H)(\mu_2\text{-}PPh_2)(CO)_{8-n}L_n]$ ($n = 1$, 2 or 4) (table 2). No simple 1:1 addition products corresponding to fission of the hydrogen-bridged metal–metal bond could be identified.

With alkynes, $RC\vdots CR'$, a reaction takes place under thermal or photolytic conditions, which is in many ways analogous to that of $[Os_3H_2(CO)_{10}]$ with acetylene. The complexes

TABLE 2. SUBSTITUTION PRODUCTS DERIVED FROM $[Mn_2(\mu_2\text{-}H)(\mu_2\text{-}PPh_2)(CO)_8]$

compound	$\nu(CO)/cm^{-1}$†	¹H n.m.r.‡
$[Mn_2(\mu_2\text{-}H)(\mu_2\text{-}PPh_2)(CNBu^t)(CO)_7]$	2074m, 2020s, 2000s, 1974s, 1953s, 1947m, 1927m	7.46(m, 10H, Ph), 4.29(s, 9H, Bu^t), −15.06(d, $^2J_{PH}$ = 34.9 Hz, 1H, MnH)
$[Mn_2(\mu_2\text{-}H)(\mu_2\text{-}PPh_2)(PPh_3)(CO)_7]$	2073m, 2023m, 1989s, 1953s, 1940m, 1917m	7.78(m, 25H, Ph), −15.82 (t, $^2J_{PH}$ = 30.8 Hz, 1H, MnH)
$[Mn_2(\mu_2\text{-}H)(\mu\text{-}PPh_2)P(OMe)_3(CO)_7]$	2075m, 2034m, 1992s, 1950s, 1926m	7.70(m, 10H, Ph), 3.89(d, $^3J_{PH}$ = 12.0, 9H, OMe), −16.46(t, $^2J_{PH}$ = 30.6, 1H, MnH)
$[Mn_2(\mu_2\text{-}H)(\mu_2\text{-}PPh_2)(CNBu^t)_2(CO)_6]$	2016m, 2004s, 1950s, 1944w, 1924s	7.67(m, 10H, Ph), 0.99(s, 18H, Bu^t)′−16.38(d, $^2J_{PH}$ = 36.6 Hz, 1H, MnH)
$[Mn_2(\mu_2\text{-}H)(\mu_2\text{-}PPh_2)\{(EtO)_2POP(OEt)_2\}(CO)_6]$	2034s, 2003s, 1959m, 1925s	7.89(m, 10H, Ph), 4.11(m, 20H, Et)′−17.11(q, $^2J_{PH}$ = 28.9 Hz, 1H, MnH)
$[Mn_2(\mu_2\text{-}H)(\mu_2\text{-}PPh_2)\{(EtO)_2POP(OEt)_2\}_2(CO)_4]$	2075m, 2029m, 1989vs, 1945s, 1929sh	7.69(m, 10H, Ph), 4.22(m, 40H, Et), −16.72(m, 1H, MnH)

† Recorded in cyclohexane solution.
‡ In millionths relative to $SiMe_4$ and recorded in CD_2Cl_2 solution at room temperature.

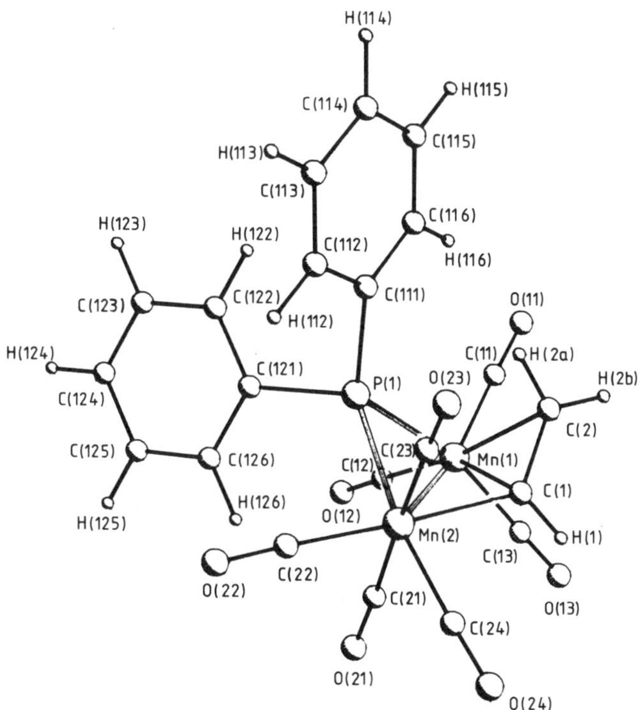

FIGURE 4. The molecular structure of $[Mn_2(\mu_2\text{-}\eta^2\text{-}HC\colon CH_2)(\mu_2\text{-}PPh_2)(CO)_7]$, including the atom numbering scheme with some bond parameters: Mn(1)–Mn(2), 2.750(1) Å; Mn(1)–C(1), 2.086(3) Å; Mn(1)–P(1), 2.253(2) Å; Mn(1)–C(2), 2.263(4) Å; Mn(2)–P(1), 2.369(1) Å; Mn(2)–C(1), 2.057(6) Å; C(1)–C(2), 1.372(7) Å; angle Mn(1)–P(1)–Mn(2), 73.0(1)°. (1 Å = 10^{-10} m = 10^{-1} nm.)

$[Mn_2(\mu_2\text{-}\eta^2\text{-}RC:C(H)R')(\mu_2\text{-}PPh_2)(CO)_7]$ resulting from insertion of the alkyne into a metal–hydrogen bond are the principal products, and an X-ray analysis of the acetylene adduct (Henrick *et al.* 1982) has established its molecular structure (figure 4).

There are two alternative pathways that can be conceived for the insertion reactions we have described, and these are shown in scheme 1. Route (*a*) involves the initial loss of a CO ligand to give the unsaturated species $[Mn_2H(CO)_7PPh_2]$, which can then react with the

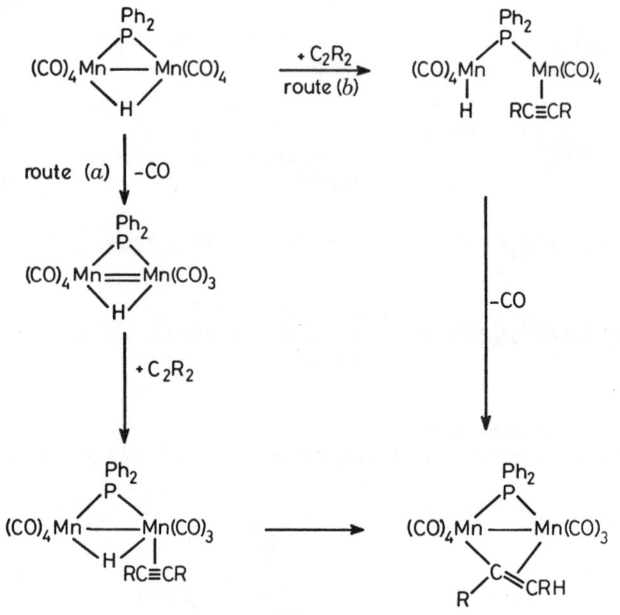

SCHEME 1

alkyne in the same manner as proposed for $[Os_2H_2(CO)_{10}]$. Route (*b*) involves the initial co-ordination of the alkyne without CO loss, which presumably involves fission of the hydrogen-bridged metal–metal bond. The results of experiments designed to differentiate between these pathways have so far proved inconclusive. Thus although the fact that the reaction proceeds under photolytic conditions suggests that CO dissociation may be involved, photolysis of **2** in THF or hydrocarbon solvents in the presence of a stream of N_2 to drive off any dissociated CO causes no change either in the infrared spectrum or in the colour of the solution, and subsequent addition of an alkyne does not give the insertion product. Whichever pathway is followed, the alkyne, once coordinated, must undergo insertion rapidly because no simple alkyne adducts (as shown in scheme 1) were isolated from reaction with any of the alkynes studied.

The reaction of electron donor ligands with these vinyl-bridged metal–metal bonded dimers has also been studied. Thus treatment of $[Mn_2(PhC:C(H)Ph)(\mu_2\text{-}PPh_2)(CO)_7]$ with $PhMe_2P$ gives the zwitterion complex $[Mn_2^-\{C(Ph)CH(Ph)P^+Me_2Ph\}(\mu_2\text{-}PPh_2)(CO)_7]$, and reaction with H^- (NaBH$_4$) gives the anionic μ_2-carbene complex $[Mn_2\{C(Ph)CH_2Ph\}(\mu_2\text{-}PPh_2)(CO)_7]^-$. The proposed structures for these complexes, based on ^1H n.m.r. data, are shown in figure 5.

Although the focus of this study has been the reactions of stabilized clusters with electron-donor ligands we have also examined the reactions of the ruthenium and manganese complexes with electrophiles: thus **1** on dissolution in trifluoroacetic acid gives the cation $[Ru_3(\mu_2\text{-}H)_3\text{-}$

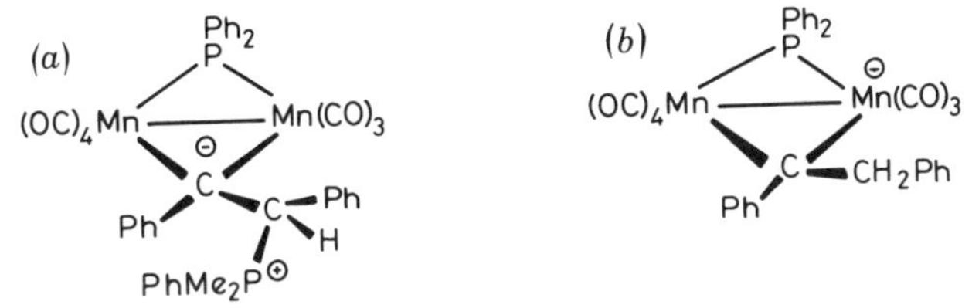

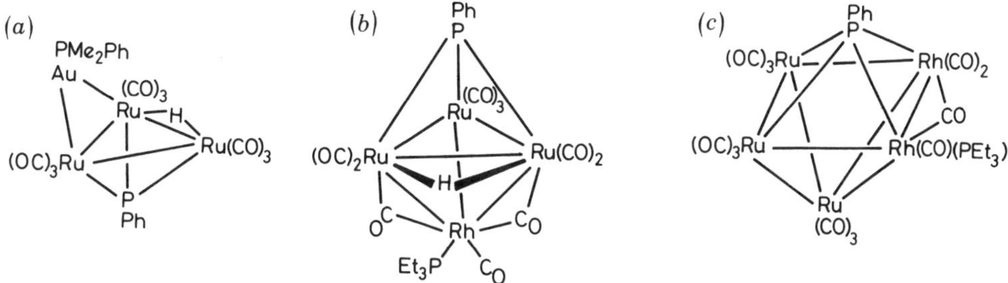

Figure 5. Proposed structure for (a) $[Mn_2^-\{C(Ph)CH(Ph)P^+Me_2Ph\}(\mu_2\text{-}PPh_2)(CO)_7]$ and (b) $[Mn_2\{C(Ph)CH_2Ph\}(\mu_2\text{-}PPh_2)(CO)_7]^-$.

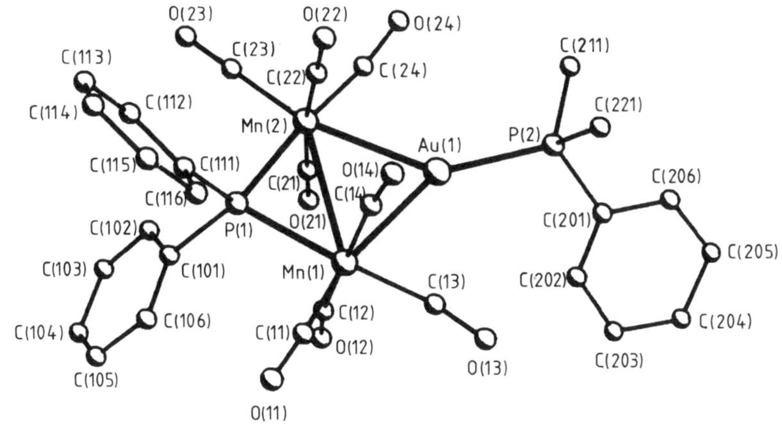

Figure 6. Some mixed-metal clusters derived from the anion $[Ru_3H(PPh)(CO)_9]^-$; (a) with $[(PhMe_2P)_2Au]^+$, (b) and (c) with $[(Et_3P)_2Rh(CO)_3]^+$.

Figure 7. The molecular structure of $[Mn_2(\mu_2\text{-}AuPMe_2Ph)(\mu_2\text{-}PPh_2)(CO)_8]$, including the atom numbering scheme with some bond parameters: Au(1)–Mn(1), 2.696(5) Å; Mn(1)–P(1), 2.297(9) Å; Au(1)–Mn(2), 2.681(5) Å; Mn(2)–P(1), 2.279(9) Å; Mn(1)–Mn(2), 3.066(7) Å; Au(1)–P(2), 2.293(7) Å; angle Mn(1)–Au(1)–Mn(2), 69.5(1)°; angle Mn(1)–P(1)–Mn(2), 84.1(3)°.

$(\mu_3\text{-}PPh)(CO)_9]^+$, with addition of a bridging hydride ligand to the third metal–metal bond. An enhanced reactivity towards electrophiles can be engendered by initial deprotonation of the complexes as described earlier for **1**. Complex **2** may be deprotonated in a similar way. In particular we have studied the reactivity of these stabilized anions towards electrophilic attack by neutral and cationic metal complexes, with a view to synthesizing mixed-metal clusters. These so-called redox-condensation reactions (Chini 1978) constitute a general method of synthesis of such clusters (Gladfelter & Geoffroy 1980) and a selection of the mixed-metal complexes containing ruthenium that we have prepared by this route and structurally characterized (Mays et al. 1982) are shown in figure 6. It is possible to substitute edge-bridging and face-bridging metal atoms for the proton removed from the starting complex. The second proton can also be substituted by further deprotonation of the mixed-metal species and subsequent reaction with an additional quantity of a cationic metal complex. With the rhodium complex $[(Et_3P)_2Rh(CO)_3]^+$ as the cation, however, a skeletal rearrangement takes place on substitution of the second proton (figure 6c). Figure 7 shows the molecular structure of a manganese–gold complex obtained from reaction of the dimanganese anion with $[(PhMe_2P)_2Au]^+$ (Iggo *et al.* 1982). With the exception of the triruthenium–dirhodium species, the complexes shown in figure 6 are obtained in near quantitative yield based on the initial stabilized neutral cluster, and studies of their reactivity towards, for example, organic substrates are therefore also feasible.

We thank Dr P. R. Raithby and Dr K. Henrick for carrying out all the X-ray analyses whose results have been described in this paper.

REFERENCES

Adams, R. D. & Golembeski, N. M. 1979 *J. Am. chem. Soc.* **101**, 2579–2587.
Adams, R. D. & Katahira, D. A. 1982 *Organometallics* **1**, 53–59.
Andrews, M. A., Kirtley, S. W. & Kaesz, H. D. 1977 *Inorg. Chem.* **16**, 1556–1561.
Banford, J., Dawoodi, Z., Mays, M. J. & Henrick, K. 1982 *J. chem. Soc. chem. Commun.*, pp. 554–556.
Chini, P., Longoni, G. & Albano, V. G. 1977 In *Advances in organometallic chemistry* (ed. F. G. A. Stone & R. West), vol. 14, pp. 285–341. New York and London: Academic Press.
Dawoodi, Z., Mays, M. J. & Henrick, K. 1982 *J. chem. Soc. chem. Commun.*, pp. 626–698.
Deeming, A. J. & Hasso, S. 1975 *J. organometall. Chem.* **88**, C21–C23.
Deeming, A. J. & Hasso, S. 1976 *J. organometall. Chem.* **114**, 313–324.
Fahmy, R., King, K., Rosenberg, E., Tiripicchio, A. & Camellini, M. T. 1980 *J. Am. chem. Soc.* **102**, 3626–3628.
Gladfelter, W. L. & Geoffroy, G. L. 1980 In *Advances in organometallic chemistry* (ed. F. G. A. Stone & R. West), vol. 18, pp. 207–273. New York and London: Academic Press.
Green, M. L. H. & Moelwyn-Hughes, J. T. 1962 *Z. Naturf.* **17b**, 783–784.
Hayter, R. G. 1964 *J. Am. chem. Soc.* **86**, 823–828.
Henrick, K., Iggo, J. A., Mays, M. J. & Raithby, P. R. 1982 *J. chem. Soc. Dalton Trans.* (In the press.)
Huttner, G., Schneider, J., Muller, H. D., Mohr, G., Seyer, J. V. & Wohlfart, L. 1979 *Angew. Chem. int. Edn.* **18**, 76–78.
Iggo, J. A., Mays, M. J., Henrick, K. & Raithby, P. R. 1982 (In preparation.)
Iwasaki, F., Mays, M. J., Raithby, P. R., Taylor, P. L. & Wheatley, P. J. 1981 *J. organometall. Chem.* **213**, 185–206.
Keister, J. B. & Shapley, J. R. 1975 *J. organometall. Chem.* **85**, C29–C31.
Knox, S. A. R., Koepke, J. W., Andrews, M. A. & Kaesz, H. D. 1975 *J. Am. chem. Soc.* **97**, 3942–3947.
Mays, M. J., Prest, D. W. & Raithby, P. R. 1980 *J. chem. Soc. chem. Commun.*, pp. 171–173.
Mays, M. J., Raithby, P. R., Taylor, P. L. & Henrick, K. 1982 *J. organometall. Chem.* **224**, C45–C48.
Shapley, J. R., Keister, J. B., Churchill, M. R. & DeBoer, B. G. 1975 *J. Am. chem. Soc.* **97**, 4145–4146.

Phil. Trans. R. Soc. Lond. A **308**, 35–45 (1982)   [35]
Printed in Great Britain

Chemistry of organocopper clusters

By J. G. Noltes

Institute of Applied Chemistry, TNO, P.O. Box 5009, 3502 JA Utrecht, The Netherlands

Thermally stable ($T_{dec} > 150\ ^\circ$C), hydrocarbon-soluble organocopper compounds have been isolated and structurally characterized. The examples discussed include mixed cluster compounds containing different organic ligands (e.g. $Cu_4(aryl)_2(alkenyl)_2$, $Cu_6(aryl)_4(alkynyl)_2$) or cluster compounds in which part of the Cu atoms have been replaced by other monovalent metals M such as Li or Au (e.g. $Cu_2M_2Ar_4$, $Cu_4M_2 Ar_4X_2$).

Organocopper compounds have polynuclear structures consisting of a metal core to which organo ligands are bound via C(1) to two metal atoms by a two-electron three-centre bond. As shown by X-ray crystallography the geometry of the $Cu_2C(1)$ moiety in organocopper clusters varies little with the nature of the bridging organo ligand (alkyl, alkenyl, alkynyl or aryl).

The C(1) atom of asymmetrically substituted aryl groups that bridge unlike metal atoms is a centre of chirality. Rotation of the aryl group around the C(1) ... C(4) axis causes a continuous inversion of configuration at C(1). Dynamic ^{1}H and ^{13}C n.m.r. studies have confirmed that such rotation actually occurs in solution for $2\text{-}Me_2NCH_2C_6H_4$–metal compounds $Ar_4M_2Li_2$ (M = Cu, Ag or Au). The prochiral methylene group offers a probe for monitoring the configuration at C(1).

The products formed in the reactions of organocopper clusters are determined to a large extent by the nature of the central copper core and by the way in which the faces of the core are occupied by the ligands. This is illustrated by the fully selective formation of the cross-coupling product observed upon thermolysis of an equimolar mixture of arylcopper and alkynylcopper compounds. This reaction proceeds via a mixed aryl–alkynyl Cu_6 cluster. Specific biaryl formation upon reaction of arylcopper compounds with a catalytic amount of copper triflate (CuOTf) likewise occurs intramolecularly in mixed aryl–triflate Cu_n clusters.

1. Introduction

In recent years organocopper compounds have gained an important role in organic synthesis as intermediates in carbon–carbon bond-forming reactions. The extreme versatility of organocopper reagents, which often show advantages over the more traditional Grignard or organolithium reagents, has strongly stimulated interest in organocopper chemistry. The organocopper reagents applied in synthetic studies have generally been reacted with the substrate molecule without prior isolation. Studies dealing with the nature and structure of organocopper compounds have received relatively less attention. The susceptibility of the copper–carbon bond to oxidation and hydrolysis as well as the lack of solubility of most organocopper compounds in hydrocarbon solvents have thwarted many attempts at isolating pure compounds.

It is only during the past decade that a limited number of organocopper compounds have been isolated in an analytically pure form and subjected to detailed structural studies in the solid state and in solution. As a result our understanding of the nature of organocopper compounds and the pathways by which they react is now beginning to develop (for a recent review see van Koten & Noltes 1982).

In this paper attention will be given to the synthesis and structural characterization of poly-nuclear organocopper compounds containing aryl, alkenyl and alkynyl ligands. The examples discussed, which represent thermally stable, hydrocarbon-soluble compounds, are taken from research at the Institute of Applied Chemistry, Utrecht. Furthermore, the dynamic behaviour of the organo ligand in arylcopper clusters, which involves a novel type of chirality, will be discussed. Examples of extremely specific carbon–carbon bond formation reactions occurring at the surface of organocopper clusters will also be given.

2. Synthesis, structure and bonding in organocopper clusters

Simple alkylcopper compounds have low thermal stability, e.g. MeCu and EtCu decompose well below room temperature. Simple arylcopper compounds are somewhat more thermally stable, but PhCu, for example, is virtually insoluble in non-coordinating solvents. Alkenylcopper compounds like vinylcopper and *cis*- and *trans*-propenylcopper also decompose well below 0 °C.

In our work on arylcopper and alkenylcopper compounds we have followed an approach that earlier had allowed the isolation of stable aryl derivatives of transition metals. Rather than add an external ligand we have stabilized the Cu–C bond by introducing a potentially co-ordinating ligand in the aryl or alkenyl group bound to copper via a Cu–C bond in a position suitable for intramolecular coordination to occur. As internal or built-in ligands the Me_2N or Me_2NCH_2 groups have been used.

Organocopper compounds of this type are readily obtained by the reaction of the correspond-ing organolithium reagent with cuprous bromide in an ether solvent, e.g.

$$CuBr + LiC_6H_4CH_2NMe_2\text{-}2 \xrightarrow{Et_2O} CuC_6H_4CH_2NMe_2\text{-}2 + LiBr. \tag{1}$$

The arylcopper compound shown in (1) is hydrocarbon-soluble and thermally stable up to 180 °C. In benzene solution the compound occurs as a tetramer $[Cu_4(C_6H_4CH_2NMe_2\text{-}2)_4]$. An X-ray structural determination has revealed that in the solid this compound is also a tetramer (figure 1). The most remarkable feature is the presence of aryl groups bridging the edges of a butterfly-shaped Cu_4 frame with extremely short Cu–Cu distances (2.38 Å†). As a result of aryl to Cu_2 bridging and the occurrence of Cu–N intramolecular coordination the Cu atoms are tri-coordinate (van Koten & Noltes 1975 a). It is possible to replace some of the copper atoms in this type of cluster by other monovalent metal atoms such as lithium (van Koten & Noltes 1979 a) or gold (I) (van Koten & Noltes 1980) with retention of the cluster, e.g.:

$$[Cu_4(C_6H_4CH_2NMe_2\text{-}2)_4] + [Li_4(C_6H_4CH_2NMe_2\text{-}2)_4] \longrightarrow 2[Cu_2Li_2(C_6H_4CH_2NMe_2\text{-}2)_4]; \tag{2}$$

$$[Au_2Li_2(C_6H_4CH_2NMe_2\text{-}2)_4] + 2CuI \longrightarrow [Cu_2Au_2(C_6H_4CH_2NMe_2\text{-}2)_4] + 2LiI. \tag{3}$$

These compounds are thermally stable up to 170 °C and exist in benzene as discrete tetra-nuclear species. The formation of the Cu_2Li_2 cluster shown in (2) exemplifies cluster-exchange between tetranuclear clusters containing different metals. By using similar reactions the corresponding $Ar_4Ag_2Li_2$ and $Ar_4Au_2Li_2$ clusters have been obtained. ^{13}C and ^1H n.m.r. spectroscopy has revealed unambiguously (cf. observation of $J(^{107,\,109}Ag\text{–}^{13}C_1)$ and $J(^7Li\text{–}^{13}C_1)$

$$† \; 1 \; Å = 10^{-10} \; m = 10^{-1} \; nm.$$

that each of the four aryl groups bridges one of the Group IB metals and one lithium atom giving rise to an M_2Li_2 core with the two M and the two Li atoms in the *trans*-positions (figure 2). Furthermore, [1]H n.m.r. spectroscopy showed that the four $2\text{-}Me_2NCH_2$ ligands are bonded to the Li atoms. Thus the Li atoms in these compounds are four-coordinate and the Group IB metals two-coordinate (van Koten & Noltes 1979).

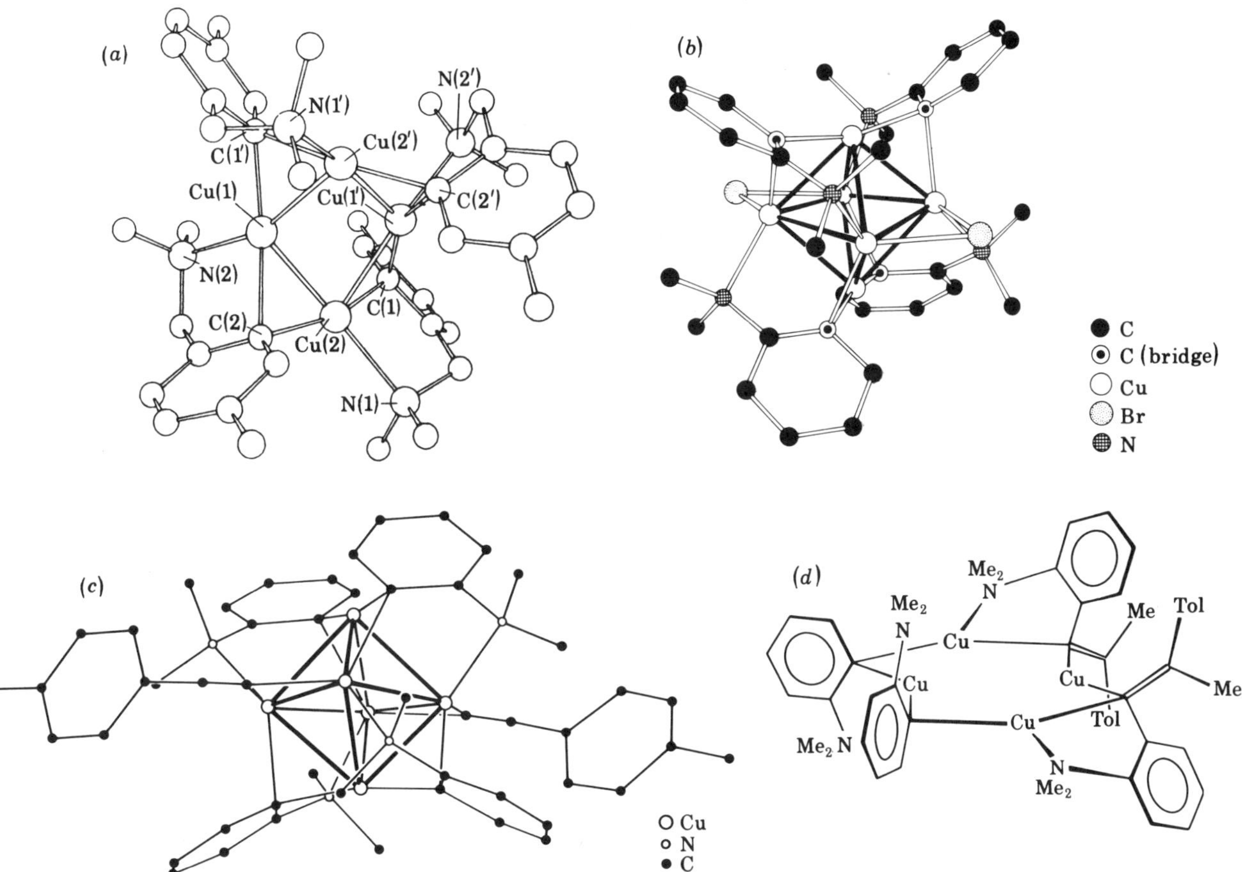

FIGURE 1. Structure of (a) $[Cu_4(C_6H_4CH_2NMe_2\text{-}2)_4]$, (b) $[Cu_6(C_6H_4NMe_2\text{-}2)_4Br_2]$, (c) $[Cu_6(C_6H_4NMe_2\text{-}2)_4(C\equiv CC_6H_4Me\text{-}4)_2]$ and (d) $[Cu_4\{(4\text{-}MeC_6H_4)MeC{=}C(C_6H_4NMe_2\text{-}2)\}_2(C_6H_4NMe_2\text{-}2)_2]$.

The presence of built-in ligands is not a prerequisite for stability in these mixed-metal clusters, e.g. tetranuclear *p*-tolyl Group IB metal–lithium compounds have been isolated. Here, the lithium atoms are three-coordinate as a result of the coordination of one molecule of diethylether (figure 2).

Simple arylcopper compounds like *o*-tolylcopper and *p*-tolylcopper are also tetranuclear in benzene solution. In this case the Cu_4 core of the molecule will be planar as a result of digonal coordination of the four Cu atoms. By using X-ray crystallographic methods the Cu_4 core of the alkylcopper compound $[Cu_4(CH_2SiMe_3)_4]$ has been shown also to have a square planar configuration (Jarvis *et al.* 1977).

In principle Ar_4M_4 and $Ar_4M_2Li_2$ compounds containing asymmetrically substituted aryl groups can exist as four unique geometric isomers depending on whether the substituents are above or below the Cu_4 or M_2Li_2 core. Figure 3 shows one of the geometric isomers of

$[Cu_4(C_6H_4Me-2)_4]$. The observation of multiplet resonances for the methyl groups, if the [1]H n.m.r. spectrum is recorded at $-60\,°C$ (Hofstee *et al.* 1978), shows that the *o*-tolylcopper tetramer can indeed be frozen out in various geometric isomers.

The structure of arylcopper compounds containing a 2-dimethylamino built-in ligand differs from that of those containing a $2\text{-Me}_2\text{NCH}_2$ substituent. In the Me_2NCH_2-substituted compounds the Me_2N group can coordinate to one of the Cu atoms bridged by the aryl group,

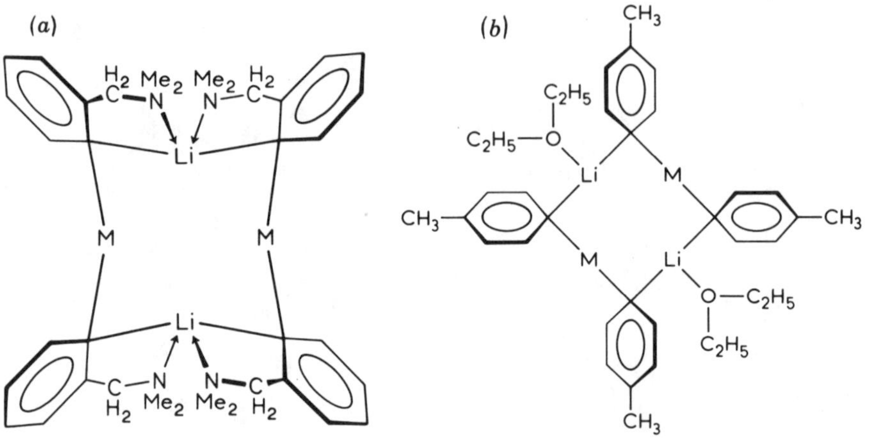

FIGURE 2. Schematic structure of (*a*) $[M_2Li_2(C_6H_4CH_2NMe_2\text{-}2)_4]$ and (*b*) $[M_2Li_2(C_6H_4Me\text{-}4)_4(Et_2O)_2]$ (M = Cu, Ag or Au).

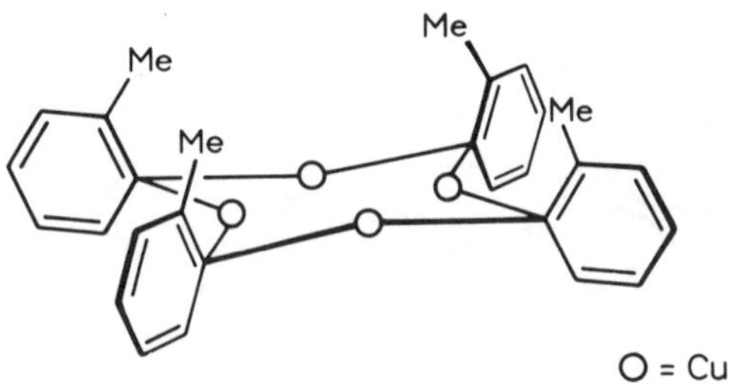

FIGURE 3. One of the geometric isomers of $[Cu_4(C_6H_4Me-2)_4]$.

thereby forming a five-membered chelate ring. The same type of coordination for the Me_2N-substituted compounds leads to a four-membered chelate ring, which is sterically unfavourable. For the latter compounds a structure in which the Me_2N group coordinates to a third Cu atom is preferred.

The reaction of $CuC_6H_4NMe_2\text{-}2$ with cuprous halides in benzene, if performed in a 2:1 molar ratio, yields bright-red, mixed aryl–halide copper clusters:

$$(4/n)\,[Cu_n(C_6H_4NMe_2\text{-}2)_n] + (2/m)[Cu_mX_m] \xrightarrow[20\,°C]{\text{benzene}} [Cu_6(C_6H_4NMe_2\text{-}2)_4X_2]. \qquad (4)$$

These compounds, which are benzene-soluble, are hexanuclear in solution. The schematic structure of the Cu_6 cluster, where X = Br, is shown in figure 1 (van Koten & Noltes 1975b). The four aryl groups bridge equatorial and apical Cu atoms of a distorted Cu_6 octahedron. The Me_2N group is coordinated to a third equatorial Ca atom. Thus each anilino ligand bridges a triangular Cu_3 face. The two Br atoms bridge *trans*-equatorial edges of the Cu_6 octahedron.

The mixed aryl–bromide Cu_6 clusters can undergo substitution of halogen with retention of the hexanuclear cluster structure. Of particular interest is the replacement of halogen by organic ligands, which gives rise to mixed organocopper cluster compounds. This type of reaction, among others, has been applied to the synthesis of mixed aryl–alkynyl Cu_6 clusters:

$$[Cu_6(C_6H_4NMe_2\text{-}2)_4Br_2] + 2LiC\equiv CC_6H_4Me\text{-}4$$
$$\longrightarrow [Cu_6(C_6H_4NMe_2\text{-}2)_4(C\equiv CC_6H_4Me\text{-}4)_2] + 2LiBr. \quad (5)$$

The same compounds can be obtained by the $2:1$ reaction of arylcopper and alkynylcopper compounds by cluster exchange:

$$(4/n)\,[Cu_n(C_6H_4NMe_2\text{-}2)_n] + (2/m)[Cu_m(C\equiv CC_6H_4Me\text{-}4)_m] \longrightarrow$$
$$[Cu_6(C_6H_4NMe_2\text{-}2)_4(C\equiv CC_6H_4Me\text{-}4)_2]. \quad (6)$$

X-ray structural determination has confirmed that the octahedral structure is retained upon ligand substitution (ten Hoedt *et al.* 1979*a*) (figure 1). The arylacetylide ligands bridge the *trans*-edges of the equatorial plane symmetrically. The acetylenic carbon atoms, which show extremely short $C\equiv C$ distances of 1.175 Å, are colinear in the equatorial plane.

The ability of the 2-$Me_2NC_6H_4$ group (Ar) to act as a tridentate ligand by spanning three metal atoms confers considerable stability on the Ar_4M_6 skeleton. A variety of hexanuclear Group IB metal compounds have been prepared in which Cu atoms in $Ar_4Cu_6X_2$ have been replaced by Ag or Au atoms (van Koten *et al.* 1977*c*):

$$(4/n)\,[Cu_nAr_n] + 2\,AgOTf \longrightarrow Ar_4Cu_4Ag_2(OTf)_2; \quad (7)$$

$$(4/n)\,[Au_nAr_n] + 4\,CuOTf \longrightarrow Ar_4Cu_4Au_2(OTf)_2 + 2\,AuOTf; \quad (8)$$

$$[Au_2Li_2Ar_4] + 4\,AgOTf \longrightarrow Ar_4Ag_4Au_2(OTf)_2 + 2\,LiOTf. \quad (9)$$

As shown in figure 1, the apical metal atoms in $Ar_4Cu_6X_2$ compounds are two-coordinate and the equatorial metal atoms three-coordinate. For monovalent Group IB metals the tendency for linear twofold coordination increases in the series $Cu^I < Ag^I < Au^I$. Indeed, in these mixed clusters the Au atoms show a distinct site preference, with the Au atoms occupying the apical positions of the M_4M_2' octahedron. This structure has been confirmed by Au–Mössbauer experiments.

In contrast to the thermally quite labile *cis*- and *trans*-propenylcopper compounds, perfectly stable ($T_{dec} > 150\,°C$) alkenylcopper compounds may be obtained if the alkenyl ligand is sterically crowded and contains a potentially coordinating ligand suitable for intramolecular coordination (ten Hoedt *et al.* 1979*b*):

$$2\,Li(\text{alkenyl}) + 2\,CuBr \longrightarrow [Cu_4(\text{alkenyl})_2Br_2] \quad (10)$$

(alkenyl $= (4\text{-}MeC_6H_4)MeC=C(C_6H_4NMe_2\text{-}2)$).

Upon the attempted preparation of alkenylcopper compounds $[Cu_n(\text{alkenyl})_n]$, mixed alkenyl–bromo Cu_4 clusters form spontaneously. Like the $Ar_4Cu_6Br_2$ clusters, the alkenyl–bromo Cu_4 compounds undergo ligand-substitution reactions with retention of the tetranuclear structure. In this way mixed alkenyl–alkynyl and mixed alkenyl–aryl clusters have been isolated (ten Hoedt *et al.* 1979*b*):

$$[Cu_4(\text{alkenyl})_2Br_2] + 2\,ArLi \longrightarrow [Cu_4(\text{alkenyl})_2(Ar)_2] + 2\,LiBr \quad (11)$$

(Ar $= C_6H_4NMe_2\text{-}2$).

The schematic structure of an (alkenyl)$_2$(aryl)$_2$Cu$_4$ cluster which has been solved by X-ray analysis (Noltes *et al.* 1982) is shown in figure 1. The two alkenyl and aryl ligands, which are in a *cis*-position, are again each bridging two Cu atoms. Only the Me$_2$N groups attached to the alkenyl ligand are coordinated, giving rise to the presence of two two-coordinate and two three-coordinate copper atoms which are in mutually *trans*-positions. As a result of the different co-ordination geometries the four Cu atoms are arranged in a rhombus pattern and not in a square as in [Cu$_4$(CH$_2$SiMe$_3$)$_4$].

The structure of the aryl–alkenylcopper cluster confirms that multicentre bonding is the preferred bonding mode of organic groups in organocopper compounds. This is true not only for alkyl, alkenyl, alkynyl and aryl groups, but also for more exotic organic ligands, e.g. (2-dimethylaminomethyl)ferrocenylcopper is tetranuclear in the solid as a result of the cyclopentadienyl C$_1$ carbon atoms bridging the edges of a Cu$_4$ square (Nesmeyanov *et al.* 1977).

FIGURE 4. Molecular orbitals involved in two-electron–three-centre ArM$_2$ bonding.

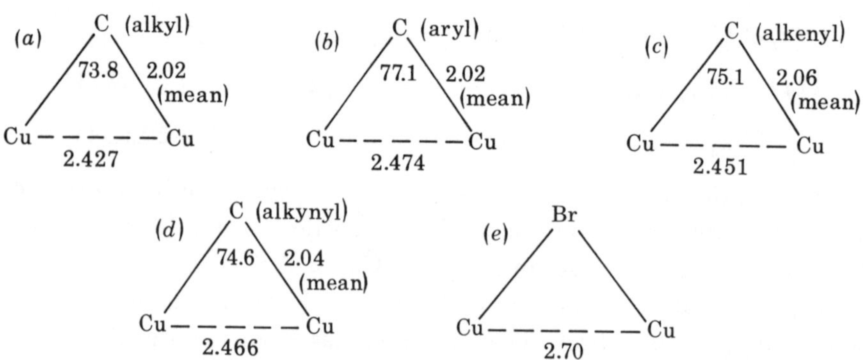

FIGURE 5. Geometry of the Cu$_2$C$_{bridge}$ moiety in some organocopper clusters: (*a*) (Me$_3$SiCH$_2$)$_4$Cu$_4$; (*b*) Ar$_2$Cu$_4$(alkenyl)$_2$; (*c*) Ar$_2$Cu$_4$(alkenyl)$_2$; (*d*) Ar$_4$Cu$_6$(alkynyl)$_2$; (*e*) Ar$_4$Cu$_6$Br$_2$. Bond lengths in ångströms, angles in degrees.

The bonding of the organic ligands in organocopper compounds can be described in terms of a simple two-electron–three-centre bonding model (figure 4). The molecular orbital that is lowest in energy results from overlap of hybrids of s and p orbitals on copper (sp or sp^2 depending on the coordination geometry) and an sp^2 orbital of the bridging aryl carbon. This molecular orbital is also bonding with respect to the two Cu atoms, and thus we speak of 'assisted' Cu–Cu bonding, as distinct from direct Cu–Cu bonding, taking place. Direct Cu–Cu bonding in organocopper clusters is considered to be unimportant on the basis of the high energy requirements for the necessary promotion of d electrons to s or p levels. This conclusion is supported by the absence of any spectroscopic evidence (i.r., Raman or $^{63,\,65}$Cu n.q.r.) for such an interaction. The second molecular orbital involves overlap of a carbon p$_z$ orbital with an antibonding combination of Cu orbitals. Bonding will be optimizal when the direction of the p$_z$ orbital is parallel to the Cu–Cu vector. Back-donation from copper to ligand likewise requires an antibonding combination of Cu orbitals. The shortest Cu–Cu distance is, of course, expected, if only the first-type molecular orbital is occupied.

The results of the various X-ray structure determinations have revealed that the geometry of the Cu_2C(bridge) moiety in multicentre bonded organocopper compounds varies little with the nature of the organo ligand (figure 5). Replacement of, for example, an alkynyl group by a bromide atom results in considerable lengthening (from 2.47 to 2.70 Å) of the bridged Cu–Cu distance. Because the bromine atom acts as a three-electron donor (four-electron–three-centre bonding), an antibonding combination of Cu orbitals is involved (figure 4), which accounts for the longer Cu_{eq}–Cu_{eq} distance in $Ar_4Cu_6Br_2$ (ten Hoedt *et al.* 1979*a*).

3. DYNAMIC BEHAVIOUR OF ARYL GROUPS IN ARYLCOPPER CLUSTERS

As shown in figure 4 the two-electron–three-centre bond has axial symmetry for the lowest energy molecular orbital. Orbital overlap is independent of the position of the aryl nucleus with respect to the Cu–Cu vector. Rotation of the aryl group around the C(1)–C(4) axis does not affect the A-type bridge bond and therefore might well occur in solution. It turns out that the aryl-bridged mixed metal clusters discussed in §2 are exactly suitable for obtaining information concerning the question whether rotation of two-electron–three-centre bonded aryl groups actually takes place in solution.

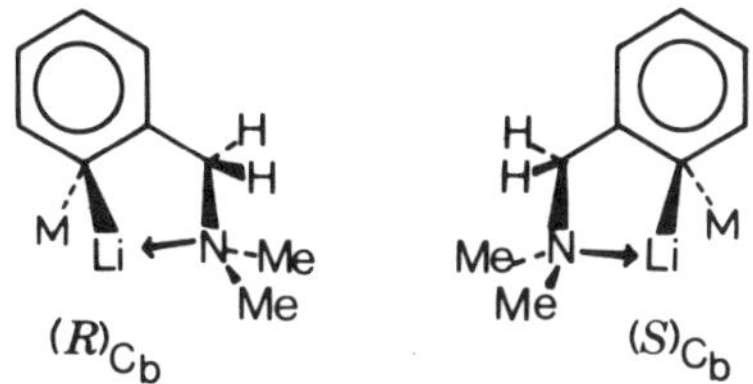

FIGURE 6. Enantiomeric pair of one rotamer conformation of the $(2\text{-}Me_2NCH_2C_6H_4)CuLi$ unit.

The bridging aromatic carbon atom in compounds $[M_2Li_2(C_6H_4CH_2NMe_2\text{-}2)_4]$ (M = Cu, Ag or Au) (see figure 2) represents a centre of chirality: (i) the aryl group bridges unlike metal atoms; (ii) the aryl group is asymmetrically substituted. The enantiomeric pair of one rotamer conformation of the $2\text{-}Me_2NCH_2C_6H_4CuLi$ unit is shown in figure 6. It should be noted that rotation of the two-electron–three-centre bonded aryl group around the C(1)–C(4) axis will cause a continuous inversion of configuration at C_{bridge}. This process at C_{bridge} does not involve bond dissociation–association processes (e.g. S_N1 or S_N2). The term fluxional chirality has been used in this connection (van Koten & Noltes 1979*b*).

The $2\text{-}Me_2NCH_2C_6H_4MLi$ unit contains two prochiral centres. Therefore ^1H and ^{13}C n.m.r. spectroscopy allow the detection of whether C_{bridge} is a stable or fluxional chiral centre.

The ^1H n.m.r. resonance patterns for the methylene and methylamino protons in $[Cu_2Li_2(C_6H_4CH_2NMe\text{-}2)_4]$ are temperature dependent. At $-60\ °C$ an AB pattern is observed for the CH_2N protons and two singlets are observed for the NCH_3 protons. At room temperature the latter two singlets coalesce to one singlet. The AB pattern for the CH_2N protons is present up to $80\ °C$. The coalescence temperatures for the two patterns in conjunction with the $\Delta\delta$ values (100 MHz; CH_2, $T_c \approx 90\ °C$, $\Delta\delta = 1.6$–1.9×10^{-6}. CH_3, $T_c \approx 5\ °C$, $\Delta\delta = 0.7 \times 10^{-6}$) indicate that two different processes are responsible for the dynamic behaviour of the NCH_3 and CH_2N protons. The scheme shown in figure 7, which involves (i) Li–N dissociation and (ii) rotation of

42   J. G. NOLTES

the aryl group around the C(1)–C(4) axis, accounts for the observed spectra. The aryl group is anchored in a fixed position if the Me_2N group is coordinated to the Li atom. Rotation of the aryl group is not possible and C_{bridge} is a stable chiral centre. Li–N coordination renders the N atom a stable prochiral centre because inversion at N is blocked. Both the CH_2 protons and the NMe_2 groups are therefore diastereotopic, as is indeed observed at -60 °C. Coalescence of the NMe_2 singlets is explained by a process involving Li–N bond dissociation, inversion at N and concomitant CH_2–N bond rotation followed by coordination. The coupled inversion and C–N bond rotation processes are fast over the whole temperature range studied. Both are low-energy processes with barriers amounting to 25 kJ mol^{-1}. The observation of an AB pattern for

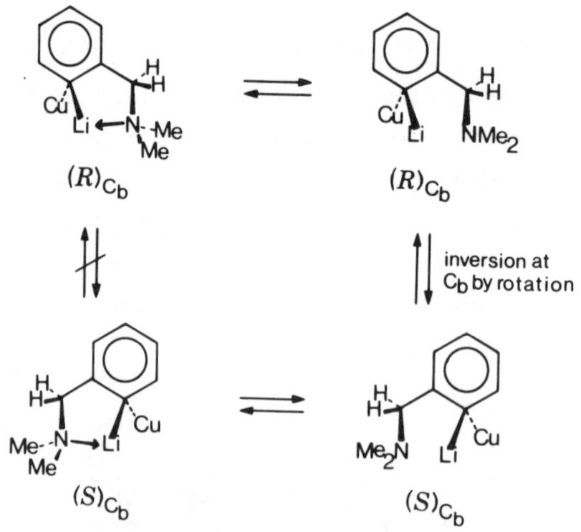

FIGURE 7. Li–N dissociation – aryl rotation process accounting for the dynamic n.m.r. spectra of
$[M_2Li_2(C_6H_4CH_2NMe-2)_4]$.

the NCH_2 protons up to 80 °C indicates that inversion of configuration at C_{bridge} does not occur on the n.m.r. timescale. It is only at 90 °C that the CH_2 protons become isochronous and this can only be explained by rapid inversion of configuration at C_{bridge}, i.e. by rotation of the aryl group around the C(1)–C(4) axis (van Koten & Noltes 1979b).

According to one of the longstanding axioms of organic chemistry, racemization of chiral carbon compounds does not occur without the breaking of a bond at the chiral carbon atom (Hückel 1931). This axiom does not necessarily hold for multicentre bonded organometallic clusters, as shown by the results presented here. That indeed no bonds are broken in the $M_2Li_2Ar_4$ clusters is confirmed by the observation that, e.g., $J(^{107,\,109}Ag–^{13}C_{br})$ and $J(^7Li–^{13}C_{br})$ remain unaltered over the temperature range -60 °C to $+90$ °C.

Introduction in the aryl group of a second centre of chirality of which the configuration cannot invert results in two diastereomeric forms of the Ar_2CuLi unit. Figure 8 shows the two diastereomeric forms of the Ar_2CuLi unit obtained on replacing the Me_2NCH_2 groups in $[Cu_2Li_2(C_6H_4CH_2NMe_2)_4]$ by S-2-Me_2NC^*HMe (Ar$'$) groups. In principle, this technique of chiral labelling allows the detection of the stereochemistry of the C_{bridge} atoms in (S)-Ar$'_4M_2Li_2$ clusters, because the diastereomeric units will display different n.m.r. chemical shifts, e.g. the observation at 27 °C of only one resonance pattern for the various protons in the ^1H n.m.r. spectrum of (S)-Ar$'_4Au_2Li_2$ simply allows the conclusion that all four C_{bridge} atoms have the same configuration. At this temperature aryl group rotation is blocked and hence C_{bridge} is a stable chiral centre. Therefore, two patterns for each signal would have been present if C_{bridge} atoms with both S and R configuration had been present. The presence of the additional Me group in the built-in ligand apparently results in the fully stereoselective formation of only one $(S)\text{-}Ar_4'Au_2Li_2$ stereoisomer in which all four C_{bridge} atoms have the same configuration, either S or R (van Koten & Noltes 1979b).

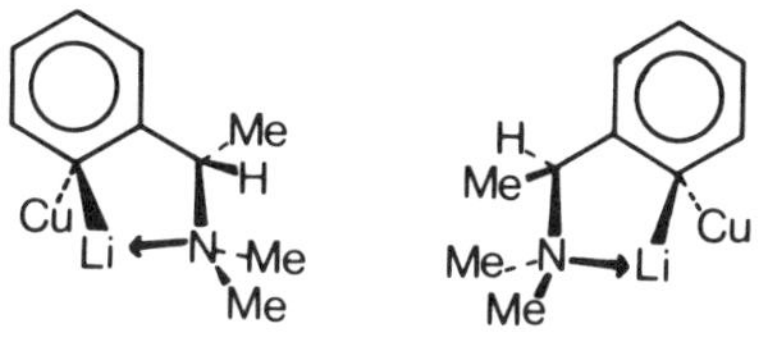

FIGURE 8. Diastereomeric pair of one rotamer conformation of the $(S\text{-}2\text{-}Me_2NCHMeC_6H_4)CuLi$ unit.

4. SELECTIVE C–C BOND FORMATION AT THE SURFACE OF COPPER CLUSTERS

Organocopper compounds both in the solid and in solution possess well-defined polynuclear structures. It is to be expected that product formation in reactions of organocopper compounds will be determined to a large extent by the nature of the central copper core and by the way in which the faces of the copper core are occupied by the ligands.

The selective cross-coupling observed upon thermolysis of an equimolar mixture of $CuC_6H_4NMe_2\text{-}2$ (CuAr) and a copper arylacetylide ($CuC{\equiv}CAr'$) is a case in point. In principle, symmetric (formation of ArAr and $Ar'C{\equiv}C\text{-}C{\equiv}CAr'$) as well as asymmetric (formation of $ArC{\equiv}CAr'$) coupling might be expected to take place. If free radicals are involved, the formation of arene (ArH) and alkyne ($Ar'C{\equiv}CH$) must be expected. In practice, only one product, the cross-coupling product $Ar{\equiv}CAr'$, is observed. This result can be rationalized on the basis of the molecular architecture of mixed arylcopper–alkynylcopper clusters occurring as intermediates in these reactions (van Koten $et\ al.$ 1977b; ten Hoedt $et\ al.$ 1979a).

The thermolysis of an equimolar mixture of CuAr and $CuC{\equiv}CAr'$ (equation (16)) proceeds in several discrete steps (equations (12)–(15)):

$$4CuAr + 2CuC{\equiv}CAr' \longrightarrow [Cu_6Ar_4(C{\equiv}CAr')_2]; \tag{12}$$

$$[Cu_6Ar_4(C{\equiv}CAr')_2] \longrightarrow ArC{\equiv}CAr' + [Cu_4^I Cu_2^0 Ar_3(C{\equiv}CAr')]; \tag{13}$$

$$[Cu_4^I Cu_2^0 Ar_3(C{\equiv}CAr')] \longrightarrow 2CuAr + [Cu_2^I Cu_2^0 Ar(C{\equiv}CAr')]; \tag{14}$$

$$[Cu_2^I Cu_2^0 Ar(C{\equiv}CAr')] \longrightarrow 4Cu^0 + ArC{\equiv}CAr'; \tag{15}$$

$$2CuAr + 2CuC{\equiv}Car' \longrightarrow 4Cu^0 + 2ArC{\equiv}CAr'. \tag{16}$$

The first and crucial step is the formation of the mixed aryl–alkynyl Cu_6 cluster from the two organocopper compounds (equation (12)). The structure of an example of a $[Cu_6Ar_4(C{\equiv}CAr')_2]$ cluster that has been solved by X-ray analysis is shown in figure 1. In this structure only triangular faces occupied by one aryl and one alkynyl ligand are present. Cu_3 faces occupied by two aryl or two alkynyl ligands are absent. This specific arrangement is responsible for the formation of asymmetric $Ar(C{\equiv}CR)Cu^{II}$ centres by an intra-aggregate valence disproportionation from which $ArC{\equiv}CAr'$ is formed by reductive elimination (equation (13)). The Cu_3 faces act as a template. The resulting copper species undergoes cluster reorganization (equation (14)) with the CuAr generated being recycled in step 1 (equation (12)). The resulting cluster again undergoes reductive coupling (equation (15)). Summation of (12)–(15) shows the actually observed stoichiometry of the thermolysis (equation (16)).

A second example of highly selective C–C coupling on a Cu_n cluster surface concerns the formation of biaryls through the interaction of arylcopper clusters with copper triflate (CuOTf). These reactions illustrate the strong influence of the nature of the counteranion on the behaviour of organocopper clusters (van Koten *et al.* 1977*b*).

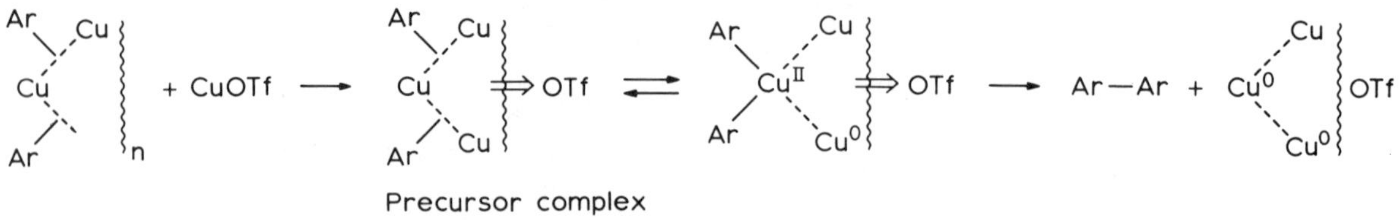

FIGURE 9. Mechanism for CuOTf-catalysed pairwise release of aryl groups from arylcopper clusters.

The reaction of $[Cu_4(C_6H_4CH_2NMe_2\text{-}2)_4]$ with a stoichiometric quantity of CuOTf leads to quantitative formation of $(2\text{-}Me_2NCH_2C_6H_4-)_2$. Similarly, $[Cu_6(C_6H_4NMe_2\text{-}2)_4Br_2]$ quantitatively yields $(2\text{-}Me_2NC_6H_4-)_2$. This Cu_6 cluster is perfectly stable in the presence of excess of CuBr. Both reactions require stoichiometric amounts of CuOTf, because the biaryls formed each coordinate to two CuOTf molecules, which are thereby blocked for further reaction. Indeed, the absence of internal ligand in the arylcopper cluster renders the reaction catalytic in CuOTf, e.g. the reaction of $[Cu_4(C_6H_4Me\text{-}4)_4]$ with a catalytic amount of CuOTf in benzene at 20 °C yields p,p'-bitolyl in 100 % yield.

In these reactions, products arising from H abstraction are not formed. This observation excludes pathways involving free radicals. The clean, quantitative decomposition points to an intramolecular pathway leading to pairwise release of aryl groups. It is proposed that these reactions proceed via an arylcopper–CuOTf precursor complex formed by extension of the copper core of the parent arylcopper cluster with one or more copper atoms of copper triflate. A representative example of this type of cluster with OTf counteranions that is sufficiently stable to be isolated and characterized is formed in this reaction:

$$(4/n)\ [Cu_n(C_6H_4NMe_2\text{-}2)_n] + 2CuOTf \longrightarrow [Cu_6(C_6H_4NMe_2\text{-}2)_4OTf_2]. \qquad (17)$$

The driving force for the coupling reaction is charge transfer in the precursor complex $[Cu_{n+m}Ar_n OTf_m]$ from the $Ar_n Cu_{n+m}$ skeleton to the strongly electron-attracting OTf ligands. The electron density in the Cu_2–C_{br} regions of the cluster and thus the kinetic stability of the Cu_2–C_{br} bonds is thereby reduced. The bond-weakening effect of electron transfer away from the arylcopper cluster framework has several precedents. The formation of ArAr and ArX in the reaction of $[Cu_4Ar_4]$ with Cu^{II} halides is one example (van Koten & Noltes 1975*c*). Such reactions proceed via an inner-sphere encounter complex with electron transfer taking place via a Cu^I–X–Cu^{II} bridge. A second example is provided by the mass spectral fragmentation pattern of $[Cu_4Ar_4]$ clusters. The parent ion $[Cu_4Ar_4]^{+\cdot}$ undergoes fragmentation exclusively by loss of a radical $Ar^\cdot$ indicating that an electron has been removed from the Cu_2C_{br} molecular orbital resulting in bond weakening rather than from the Cu_4 core of the molecule (van Koten & Noltes 1975*c*).

A mechanism for the pairwise release of aryl groups from arylcopper clusters catalysed by CuOTf is shown in figure 9. A mechanism involving intracluster valence disproportionation followed by reductive coupling best accounts for the large influence of the counteranion on the occurrence of coupling (OTf against Br). The strong electron-acceptor properties of the hard OTf ligand will favour the Cu^{II} oxidation state, whereas the softer electron-donating halide ions will favour the Cu^{I} state.

It is a great pleasure to acknowledge the major contributions made by Dr G. van Koten to the research reported here.

REFERENCES

ten Hoedt, R. W. M., Noltes, J. G., van Koten, G. & Spek, A. L. 1979*a* *J. chem. Soc. Dalton Trans.*, pp. 1800–1806.

ten Hoedt, R. W. M., van Koten, G. & Noltes, J. G. 1979*b* *J. organometall. Chem.* **179**, 227–240.

Hofstee, H. K., Boersma, J. & van der Kerk, G. J. M. 1978 *J. organometall. Chem.* **144**, 255–261.

Hückel, W. 1931 *Theoretische Grundlagen der organischen Chemie*, vol. 1, p. 252. Leipzig: Akademische Verlag-gesellschaft.

Jarvis, J. A. J., Pearce, R. & Lappert, M. F. 1977 *J. chem. Soc. Dalton Trans.*, pp. 999–1003.

van Koten, G., ten Hoedt, R. W. M. & Noltes, J. G. 1977*a* *J. org. Chem.* **42**, 2705–2711.

van Koten, G., Jastrzebski, J. T. B. H. & Noltes, J. G. 1977*b* *J. org. Chem.* **42**, 2047–2053.

van Koten, G., Jastrzebski, J. T. B. H. & Noltes, J. G. 1977*c* *Inorg. Chem.* **16**, 1782–1787.

van Koten, G. & Noltes, J. G. 1975*a* *J. organometall. Chem.* **84**, 129–138.

van Koten, G. & Noltes, J. G. 1975*b* *J. organometall. Chem.* **102**, 551–563.

van Koten, G. & Noltes, J. G. 1975*c* *J. organometall. Chem.* **84**, 419–429.

van Koten, G. & Noltes, J. G. 1979*a* *J. organometall. Chem.* **174**, 367–387.

van Koten, G. & Noltes, J. G. 1979*b* *J. Am. chem. Soc.* **101**, 6593–6599.

van Koten, G. & Noltes, J. G. 1982 In *Comprehensive organometallic chemistry* (ed. G. Wilkinson, F. G. A. Stone & E. W. Abel), vol. 2, pp. 709–763. Oxford: Pergamon Press.

van Koten, G., Schaap, C. A., Jastrzebski, J. T. B. H. & Noltes, J. G. 1980 *J. organometall. Chem.* **186**, 427–445.

Nesmeyanov, A. N., Struchkov, Yu. T., Sedova, N. N., Andrianov, V. G., Volgin, Yu. V. & Sazonova, V. A. 1977 *J. organometall. Chem.* **137**, 217–221.

Noltes, J. G., ten Hoedt, R. W. M., van Koten, G. & Spek, A. L. 1982 *J. organometall. Chem.* **225**, 365–376.

Phil. Trans. R. Soc. Lond. A **308**, 47–57 (1982)   [47]

Printed in Great Britain

Iron, cobalt and nickel carbide–carbonyl clusters by CO scission

By G. Longoni, A. Ceriotti, R. Della Pergola, M. Manassero, M. Perego, G. Piro and M. Sansoni

Istituto di Chimica Generale dell' Università e Centro del CNR per lo studio della sintesi e della struttura dei composti dei metalli di transizione, via G. Venezian 21, 20133 Milano, Italy

A new approach to the synthesis in good yields of known cobalt and iron carbide–carbonyl clusters by CO cleavage in mild conditions is reported. Cleavage of CO results from attaching an acetyl or benzoyl carbocation to the oxygen atom, and by transfer of electrons from an external source. This synthetic approach to carbide molecular clusters may be of some significance with respect to the formation of carbide atoms on to metal crystallites.

Attempts to synthesize nickel carbide clusters with the same approach have only been partly successful. The new $[Ni_9C(CO)_{17}]^{2-}$ and $[Ni_8C(CO)_{16}]^{2-}$ have been obtained more conveniently from the reaction of $[Ni_6(CO)_{12}]^{2-}$ with CCl_4. The related reaction of $[Ni_6(CO)_{12}]^{2-}$ with $Co_3(CO)_9CCl$ results in the formation of the mixed-metal carbide cluster $[Co_3Ni_9C(CO)_{20}]^{3-}$. This compound is degraded under a carbon monoxide and hydrogen mixture (25 °C, 1 atm) to $Ni(CO)_4$, $[Co(CO)_4]^-$ and ethane. Intermediate formation of $[Co_3Ni_7(C-C)(CO)_{15}]^{3-}$, in which the two carbide atoms show an interatomic separation of 1.43 Å, or of a related species, would provide a possible pathway for C–C bond formation.

1. Introduction

Cluster chemistry has often provided models for individual steps that may be involved as transient intermediates in a catalytic sequence. For instance, the recent characterization of a series of hexanuclear clusters of the cobalt subgroup (e.g. $[M_6(CO)_{15}H]^-$ (M = Co, Rh), $[Rh_6(CO)_{15}(C(=O)-X)]^-$ (X = R, OR, NHR) and $[Rh_6(CO)_{14}(\eta_3-C_3H_4)]^-$ (Chini 1980)) has shown that also a higher nuclearity cluster can at least carry the functional groups that are presumably required in oxo-synthesis. Thus the hydroformylation of olefins catalysed by cobalt carbonyl derivatives is commonly accepted to occur through a sequence of steps such as (Pino *et al.* 1977):

$$HCo(CO)_3 \overset{\text{olefin}}{\rightleftharpoons} R-Co(CO)_3 \overset{CO}{\rightleftharpoons} R-C(=O)-Co(CO)_3 \overset{H_2}{\longrightarrow} \text{aldehydes.}$$

Support for the above mechanism stems also from the isolation and characterization of a series of monomeric compounds such as $HCo(CO)_4$, $R-Co(CO)_4$ and $R-C(=O)-CO(CO)_4$, which are believed to be the 18-electron counterparts of the 16-electron active intermediates of the above sequence (Pino *et al.* 1977). In this connection, it may appear significant that a key cluster model such as $[Rh_6(CO)_{15}(C(=O)-R)]^-$ has been initially obtained from

$$3Rh_4(CO)_{12} + 2CH_2{=}CH-CH_3 + 2H_2O \xrightarrow[\text{H}_2\text{ (1 atm)}]{\text{THF, 25 °C}} 2[Rh_6(CO)_{15}(C(=O)-C_3H_7)]^-$$
$$+ 2H(THF)^+ + 2CO + 2CO_2, \quad (1)$$

and that reaction (1) in apolar solvents results in stoichiometric hydroformylation of olefins (Chini *et al.* 1972), and becomes catalytic on starting from $Co_2Rh_2(CO)_{12}$.

The search for other cluster models for hydroformylation originated the present work, which resulted in finding a ready CO cleavage reaction that may be of some significance for the formation of surface carbide atoms on metal crystallites.

2. SYNTHESIS OF COBALT AND IRON CARBIDE CLUSTERS BY CO SCISSION

The reaction of a carbonylanion with acyl chloride has been shown in the past to afford either mononuclear (e.g. $R–C(=O)–Co(CO)_4$) or polynuclear acyl derivatives (e.g. $[Rh_6(CO)_{15}$-$(C(=O)-R)]^-$). In attempting the synthesis of a hexanuclear cobalt acyl derivative, we investigated the reaction of $[Co_6(CO)_{15}]^{2-}$ with acetyl chloride. Unexpectedly, this reaction did not result in the formation of the $[Co_6(CO)_{15}(C(=O)-CH_3)]^-$ acyl derivative nor the corresponding

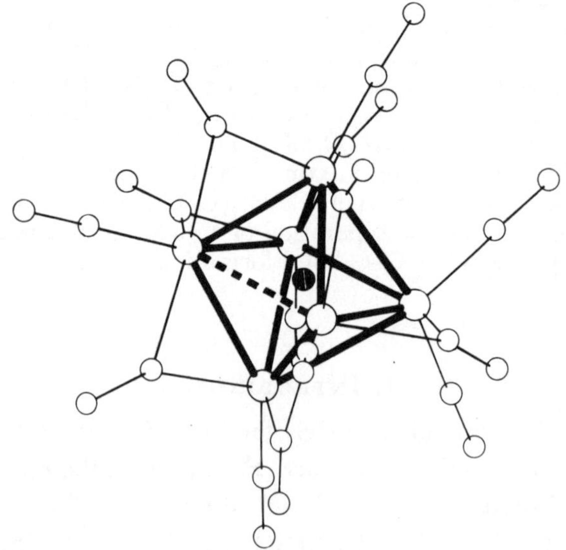

FIGURE 1. Structure of $[Co_6C(CO)_{14}]^-$ (Albano *et al.* 1980).

$[Co_6(CO)_{15}(-CH_3)]^-$ alkyl derivative, generated by decarbonylation of the former. The resulting brown compound showed an intense e.s.r. signal with $g = 2.013$, and on the basis of its spectroscopic and chemical behaviour it has been identified as the previously characterized paramagnetic $[Co_6C(CO)_{14}]^-$ carbide derivative, whose structure (Albano *et al.* 1980) is shown in figure 1.

The reaction of $[Co_6(CO)_{15}]^{2-}$ with CH_3COCl follows this apparent stoichiometry:

$$3[Co_6(CO)_{15}]^{2-} + 2CH_3COCl(+CO) \longrightarrow 2[Co_6C(CO)_{14}]^-$$
$$+ 4[Co(CO)_4]^- + 2Co^{2+} + 2Cl^- + 2CH_3COO^-. \quad (2)$$

As reported previously (Albano *et al.* 1976), the octahedral $[Co_6C(CO)_{14}]^-$ is readily converted into the prismatic $[Co_6C(CO)_{15}]^{2-}$ in the presence of basic reagents. Therefore reaction (2) can also be used to synthesize the latter direct, by a slight modification of the isolation procedure. Both $[Co_6C(CO)_{14}]^-$ and $[Co_6C(CO)_{15}]^{2-}$ have been isolated in 50–60% yields (calculated from the starting compound $K_2[Co_6(CO)_{15}]$), and reaction (2) represents an alternative route to these interstitial carbides, which have previously been synthesized by reaction of $Co_3(CO)_9CCl$ with $[Co(CO)_4]^-$ (Albano *et al.* 1974).

This unexpected result, represented by reaction (2), could have been due to ready CO scission, probably occurring with a $[Co_6(CO)_{14}(CO-C(=O)-CH_3)]^-$ transient adduct, related to the previously reported $[Fe_3(CO)_{10}(\mu_2-CO-C(=O)-CH_3)]^-$, where the acetyl carbocation is attached to the oxygen atom of a double-bridging carbonyl group (Hodali & Shriver 1979). This hypothesis led us to investigate a reaction analogous to (2) with the $[Fe_4(CO)_{13}]^{2-}$ dianion, which has recently been shown to give $[Fe_4(CO)_{12}(\mu_3-CO-CH_3)]^-$ and the corresponding $HFe_4(CO)_{12}(\mu_3-\eta_2-CO-CH_3)$ adduct, on reaction with methylating agents such as CH_3SO_3F (Holt *et al.* 1980; Dawson *et al.* 1980).

Reaction of $[Fe_4(CO)_{13}]^{2-}$ in anhydrous THF with a slight excess of acetyl chloride affords the new red-violet $[Fe_4(CO)_{12}(CO-C(=O)-CH_3)]^-$ adduct:

$$[Fe_4(CO)_{13}]^{2-} + R-COCl \longrightarrow [Fe_4(CO)_{12}(CO-C(=O)-R)]^- + Cl^- \qquad (3)$$

($R = CH_3$, C_6H_5). This compound has been isolated in *ca.* 90% yield as crude product, and its crystallization has been hampered by decomposition to the starting $[Fe_4(CO)_{13}]^{2-}$ and $[HFe_4(CO)_{13}]^-$ on standing in solution. When acetyl chloride is replaced by benzoyl chloride, reaction (3) is very slow and requires several hours for completion. The reaction is accelerated at 50–60 °C; however, long reaction time and warming favour the concurrent formation of other products. Very pure samples of $[Fe_4(CO)_{12}(CO-C(=O)-C_6H_5)]^-$ are readily available by performing reaction (3) at 18 °C and under u.v. radiation. The benzoyl adduct shows a much greater stability than the corresponding acetyl derivative, and has been recrystallized from THF-heptane as the tetrasubstituted ammonium or phosphonium salt. Both the acetyl and benzoyl adducts are supposed to be structurally related to the corresponding methyl adduct previously characterized (Holt *et al.* 1980; Dawson *et al.* 1980).

As shown in scheme 1, the acyl adducts have been converted into iron carbide derivatives by several routes. The proton-induced formation of a carbide derivative (side 1) parallels the

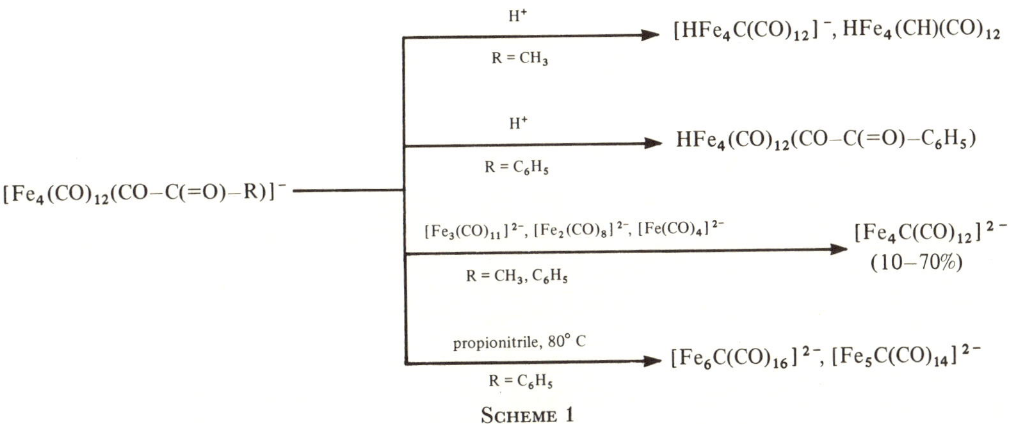

SCHEME 1

previously reported synthesis of $[HFe_4C(CO)_{12}]^-$ and $HFe_4(CH)(CO)_{12}$ from $[Fe_4(CO)_{13}]^{2-}$ and HSO_3CF_3, which probably occurs through the $HFe_4(CO)_{12}(CO-H)$ as intermediate (Holt *et al.* 1981; Whitmire & Shriver 1981). The greatest stability of the benzoyl adduct is also shown by its conversion to the corresponding $HFe_4(CO)_{12}(CO-C(=O)-C_6H_5)$ by reaction with acid (side 2).

The $[Fe_4(CO)_{12}(CO-C(=O)-R)]^-$ ($R = CH_3$, C_6H_5) derivatives are stable in the presence of excess $[Fe_4(CO)_{13}]^{2-}$; however, reaction with the more reduced $[Fe_3(CO)_{11}]^{2-}$, $[Fe_2(CO)_8]^{2-}$ or $[Fe(CO)_4]^{2-}$ gives rise to a mixture of products containing increasing amounts of the previously reported $[Fe_4C(CO)_{12}]^{2-}$ dianion (Tachikawa & Muetterties 1980). When $[Fe(CO)_4]^{2-}$ is used as reducing agent, yields up to 70–80% of $[Fe_4C(CO)_{12}]^{2-}$ have been obtained. The use of an iron carbonyl anion as reducing agent can, however, be detrimental to the selectivity of the reaction; $[Fe_4(CO)_{12}(CO-C(=O)-R)]^-$ has been almost quantitatively converted into $[Fe_4C(CO)_{12}]^{2-}$ by reduction with sodium-ketyl:

$$[Fe_4(CO)_{12}(CO-C(=O)-R)]^- + 2Na \xrightarrow{\text{THF, Ph}_2\text{CO}} [Fe_4C(CO)_{12}]^{2-} + R-COO^- + 2Na^+. \quad (4)$$

The $[Fe_4C(CO)_{12}]^{2-}$ dianion has previously been synthesized by degradation of preformed $[Fe_5C(CO)_{14}]^{2-}$, and its structure is schematically represented in figure 2 (Tachikawa & Muetterties 1980). Reaction (4) represents a convenient alternate synthesis of this interesting compound and, by subsequent protonation, for the related $[HFe_4C(CO)_{12}]^-$ and $HFe_4(CH)(CO)_{12}$ (Tachikawa & Muetterties 1980).

A mixture of $[Fe_6C(CO)_{16}]^{2-}$ (Churchill et al. 1971), $[Fe_5C(CO)_{14}]^{2-}$ (Hsieh & Mays 1972), and other unidentified products have been obtained when $[Fe_4(CO)_{12}(CO-C(=O)-C_6H_5)]^-$ is heated at 70–90 °C in propionitrile solution (side 4). Owing to difficulties encountered in

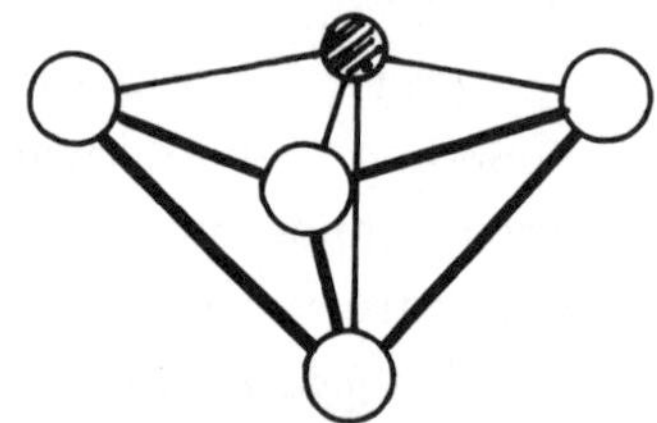

FIGURE 2. Schematic representation of the structure of $[Fe_4C(CO)_{12}]^{2-}$ (Tachikawa & Muetterties 1980).

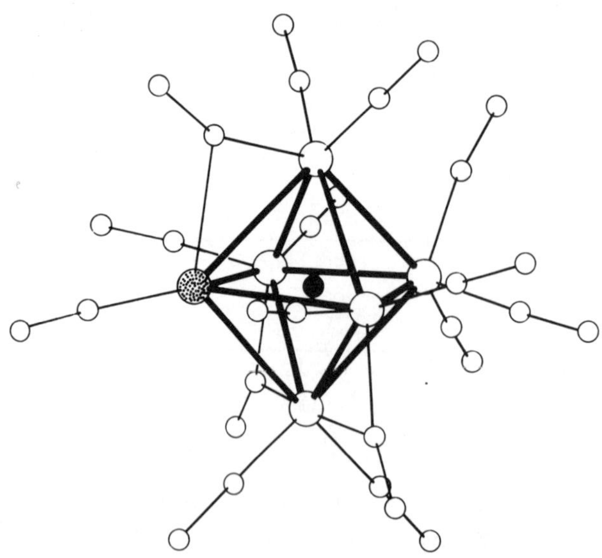

FIGURE 3. Structure of $[Fe_5PtC(CO)_{16}]^{2-}$.

the separation of this complex mixture, this direct synthetic route to hexanuclear and pentanuclear carbide clusters has not been further investigated.

Identification of the products shown in scheme 1 is based on the complete correspondence of their spectroscopic and chemical behaviour with those previously reported for these compounds. The carbide nature of the product of reaction (4) has been also confirmed by its reaction with Pt salts, and by the X-ray characterization of the resulting $[Fe_5PtC(CO)_{16}]^{2-}$. The structure of this mixed-metal carbide cluster is shown in figure 3. The isolation of a carbide derivative in side 3 of scheme 1 confirms that intermediate formation of an acyl adduct, and its ready reduction either by $[Co_6(CO)_{15}]^{2-}$ or $[Co(CO)_4]^-$, may be also at the origin of $[Co_6C(CO)_{14}]^-$ in reaction (2).

3. Synthesis of carbide nickel clusters

The ready CO cleavage reactions (2) and (4) of §2 led us to investigate the possible extension of this synthetic approach to carbide clusters of the metals of nickel subgroup. Although a Ni_3C binary bulk phase is known (Nakagura 1957), there have been no reports in the literature on the synthesis of carbide molecular clusters of nickel.

When reacting both $[Ni_5(CO)_{12}]^{2-}$ or $[Ni_6(CO)_{12}]^{2-}$ with acylchloride, we observed only oxidation to $[Ni_9(CO)_{18}]^{2-}$. Similar results have been obtained also when starting from $[Pt_6(CO)_{12}]^{2-}$. It therefore seems probable that the presence of more basic face-bridging carbonyl groups and steric crowding of the carbonyl ligands around the metal core of the cluster are important requirements to avoid side-reactions and obtain the formation of acyl adducts, in which the acyl carbocation is attached to the oxygen atom of a carbonyl group.

$$[Ni_6(CO)_{12}]^{2-} + SiCl_4 \xrightarrow[N_2]{\text{anhydrous THF}} \left| \begin{array}{l} Ni(CO)_4 \\ [Ni_9(CO)_{18}]^{2-} \\ [Ni_{12}(CO)_{21}H_2]^{2-} \\ [Ni_9C(CO)_{17}]^{2-} \\ \text{and other unidentified} \\ \text{products} \end{array} \right| \xrightarrow[CO]{\text{wet THF}} [Ni_8C(CO)_{16}]^{2-} + Ni(CO)_4$$

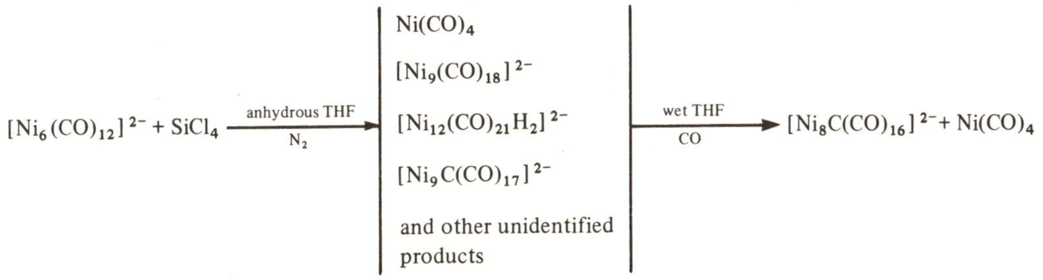

Scheme 2

To our knowledge the first genuine example of carbide synthesis by CO splitting induced by coordination of a Lewis acid to the oxygen atom of a carbonyl group was probably the fortuitous synthesis of $[Rh_6C(CO)_{15}]^{2-}$, while attempting the synthesis of a related $[Rh_6Si(CO)_{15}]^{2-}$ by reaction of $[Rh(CO)_4]^-$ with SiCl$_4$ (Chini *et al.* 1973). This result, on the basis of the known transformation of $(CH_3)_3Si–Co(CO)_4$ into $Co_3(CO)_9(\mu_3\text{-}CO\text{-}Si(CH_3)_3)$ (Ingle *et al.* 1973), may be due to a CO splitting reaction related to (2), and suggested this different approach to carbide–nickel clusters. As shown in scheme 2, the reaction of $[Ni_6(CO)_{12}]^{2-}$ in anhydrous THF with SiCl$_4$ resulted in a very complicated mixture of products. However, the subsequent degradation of this mixture under carbon monoxide in wet THF converted most of the nickel carbonyl anions into Ni(CO)$_4$, and allowed the isolation, in low yield, of a red species, which has been characterized as $[Ni_8C(CO)_{16}]^{2-}$. Despite past failures in the synthesis of carbide nickel clusters by reduction of Ni(CO)$_4$ with alkali in methanol and in the presence of small amounts of $CHCl_3$ or CCl_4, carbide nickel clusters do exist and, since the experiment shown in scheme 2, have been synthesized in good yields by the classical reaction between a preformed carbonyl anion and CCl_4 (Chini *et al.* 1974), such as the following:

$$2[Ni_6(CO)_{12}]^{2-} + CCl_4(+CO) \xrightarrow{\text{THF, N}_2} [Ni_9C(CO)_{17}]^{2-} + 2Ni(CO)_4 + Ni^{2+} + 4Cl^-. \quad (5)$$

The $[Ni_9C(CO)_{17}]^{2-}$ and $[Ni_8C(CO)_{16}]^{2-}$ dianions are related by the degradation-condensation equilibrium

$$[Ni_9C(CO)_{17}]^{2-} + 3CO \underset{N_2}{\overset{CO}{\rightleftharpoons}} [Ni_8C(CO)_{16}]^{2-} + Ni(CO)_4. \quad (6)$$

The right-hand side of this equilibrium, combined with the stability of $[Ni_8C(CO)_{16}]^{2-}$ toward further degradation under CO, is probably responsible for the isolation of this compound in the experiment represented in scheme 2.

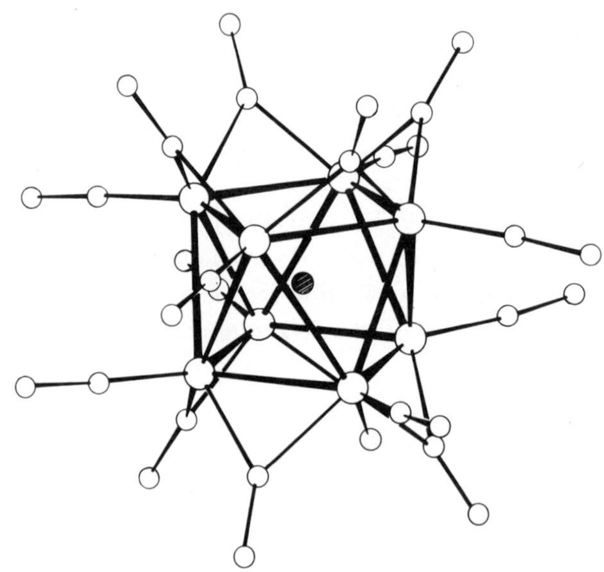

FIGURE 4. Structure of $[Ni_8C(CO)_{16}]^{2-}$.

Both these two compounds have been isolated in a crystalline state as tetrabutylammonium salts, and their X-ray structures are reported in figures 4 and 5. Although the carbide atom in the Ni_3C bulk phase has been shown to occupy octahedral cavities in both $[Ni_8C(CO)_{16}]^{2-}$ and $[Ni_9C(CO)_{17}]^{2-}$, the carbide atoms are encapsulated in a square antiprismatic cage. The Ni–Ni average bond distance in these two compounds (2.55 Å†) is intermediate between Ni–Ni bond distance in bulk Ni metal (2.50 Å) and in Ni_3C binary carbide (2.63 Å) (Nakagura 1957).

4. MIXED Co–Ni CARBIDE CLUSTERS

Reaction (7), formally analogous to reaction (5), has been used to synthesize mixed Co–Ni carbide clusters:

$$2[Ni_6(CO)_{12}]^{2-} + Co_3(CO)_9CCl \longrightarrow [Co_3Ni_9C(CO)_{20}]^{3-} + 3Ni(CO)_4 + Cl^- + CO. \quad (7)$$

$$\dagger\ 1\ \text{Å} = 10^{-10}\,\text{m} = 10^{-1}\,\text{nm}.$$

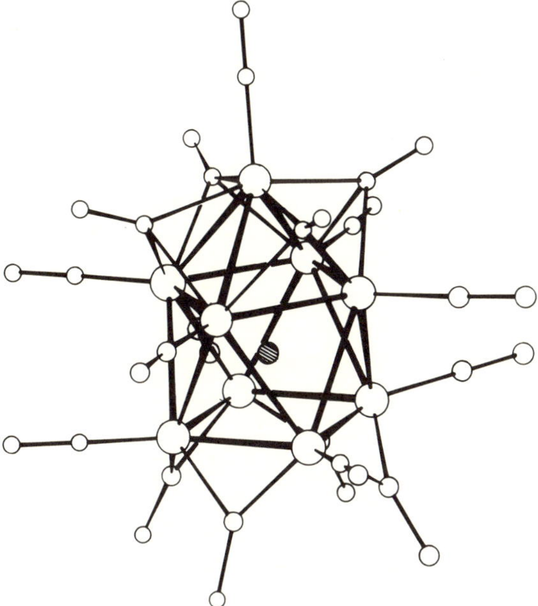

FIGURE 5. Structure of $[Ni_9C(CO)_{17}]^{2-}$.

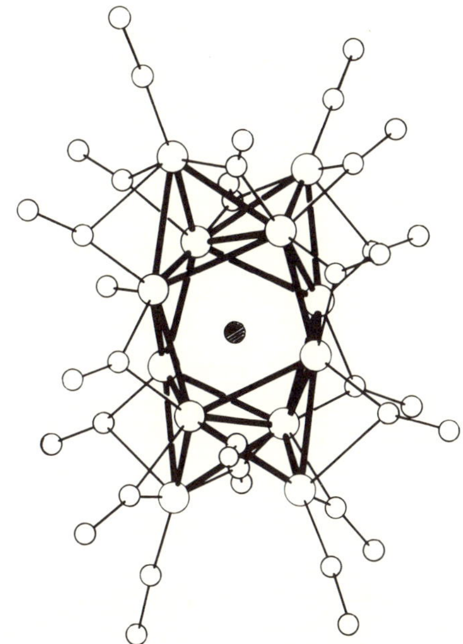

FIGURE 6. Structure of $[Co_3Ni_9C(CO)_{20}]^{3-}$.

As shown in figure 6, the structure of the $[Co_3Ni_9C(CO)_{20}]^{3-}$ is based on a tetracapped square antiprism of metal atoms. Cobalt and nickel atoms are not distinguishable and are probably randomly distributed. The most interesting chemical feature of this compound is that it differs from all the other carbide derivatives so far reported in that it is completely degraded by carbon monoxide (25 °C, 1 atm):

$$[Co_3Ni_9C(CO)_{20}]^{3-} + 28CO \longrightarrow 3[Co(CO)_4]^- + 9Ni(CO)_4 + \{C\}. \qquad (8)$$

The fate of the carbide atom in this reaction is still obscure, and the unusual unstability of $[Co_3Ni_9C(CO)_{20}]^{3-}$ under carbon monoxide is probably the consequence of the presence in this compound of the same number of cobalt atoms and free negative charges, and of the particular stability of both $[Co(CO)_4]^-$ and $Ni(CO)_4$.

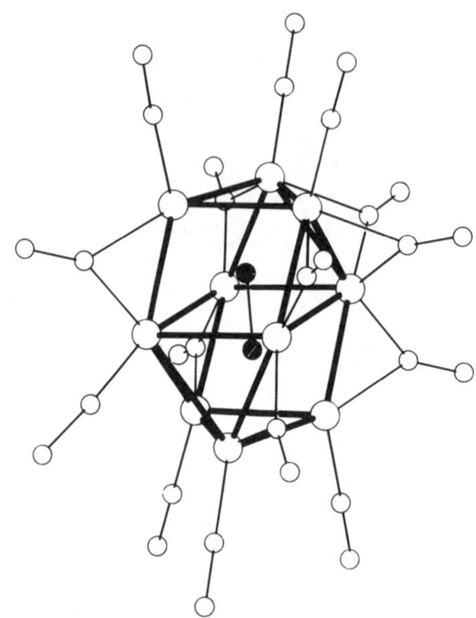

FIGURE 7. Structure of $[Co_3Ni_7C_2(CO)_{15}]^{3-}$.

When $[Co_3Ni_9C(CO)_{20}]^{3-}$ has been degraded under an atmosphere of carbon monoxide and hydrogen (1:1, 25 °C, 1 atm), the carbide atom has been almost selectively converted into ethane. A possible interpretation for the unexpected selective formation of ethane has been suggested by the subsequent isolation and characterization of the $[Co_3Ni_7C_2(CO)_{15}]^{3-}$ dicarbide, as a by-product of reaction (7). In this cluster the two interstitial carbide atoms present an interatomic separation (1.43 Å) shorter than those previously determined in $Rh_{12}C_2(CO)_{25}$ (1.48 Å) (Albano et $al.$ 1978) and $[Co_{11}C_2(CO)_{22}]^{3-}$ (1.62 Å) (Albano et $al.$ 1981). The intermediate degradation of $[Co_3Ni_9C(CO)_{20}]^{3-}$ to $[Co_3Ni_7(C-C)(CO)_{15}]^{3-}$, as represented in reactions (9) and (10), or to a structurally related species, would provide a possible pathway to C–C bond formation:

$$2[Co_3Ni_9C(CO)_{20}]^{3-} + 31CO \xrightarrow[\text{25 °C, 1 atm}]{\text{CO} + \text{H}_2} [Co_3Ni_7(C-C)(CO)_{15}]^{3-} + 3[Co(CO)_4]^- + 11Ni(CO)_4; \quad (9)$$

$$[Co_3Ni_7(C-C)(CO)_{15}]^{3-} + 25CO + 3H_2 \xrightarrow[\text{25 °C, 1 atm}]{\text{CO} + \text{H}_2} 3[Co(CO)_4]^- + 7Ni(CO)_4 + C_2H_6. \quad (10)$$

In agreement with this interpretation, reaction (10) has also been verified on starting from preformed $[Co_3Ni_7(C-C)(CO)_{15}]^{3-}$. The structure of this dicarbide is schematically shown in figure 7. The three-layer metal skeleton of this compound may be thought to derive from the rearrangement of either two interstitially occupied trigonal prismatic or octahedral moieties sharing a common edge, as shown in figure 8. Both these two undistorted arrangements may be found in binary transition metal carbides. Two trigonal prisms, condensed as shown in figure $8a$, constitute the central part of the unit cell of the Cr_3C_2 binary phase (figure 9), in which, however, the C–C interatomic separation is 1.66 Å (Toth 1971). The metal arrangement found in $[Co_3Ni_7C_2(CO)_{15}]^{3-}$ indicates the higher degree of freedom of the finite metal array of the molecular clusters, and favours both a shortening of the C–C interatomic separation and expansion of the carbide atoms in a heptahedral cage.

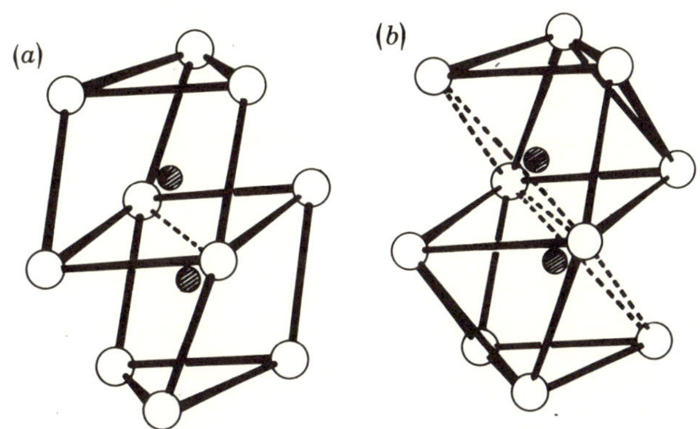

FIGURE 8. The $[Co_3Ni_7C_2(CO)_{15}]^{3-}$ metal skeleton as (a) two condensed trigonal prisms or as (b) two condensed octahedra.

This comparison represents a further confirmation of the relation between molecular carbide metal clusters and binary transition metal carbides (Tachikawa & Muetterties 1982; Chini 1980).

5. CONCLUSIONS

Several examples of intermolecular dual coordination of a metal-bound carbonyl group with a Lewis acid have been reported (Hamilton *et al.* 1981, and references therein). A significant example in cluster chemistry is given by the structure of $[Fe_4(CO)_{12}(CO-CH_3)]^-$ (Holt *et al.* 1980; Dawson *et al.* 1980), shown schematically in figure $10a$. In molecular clusters intra-molecular dual coordination of carbon monoxide has also been documented, e.g. $Mn_2(CO)_5$-$(Ph_2P-CH_2-PPh_2)_2$ (Colton *et al.* 1975), $[HFe_4(CO)_{13}]^-$ (Manassero *et al.* 1976) (figure $10b$) and $Nb_3Cp_3(CO)_7$ (Herrman *et al.* 1981).

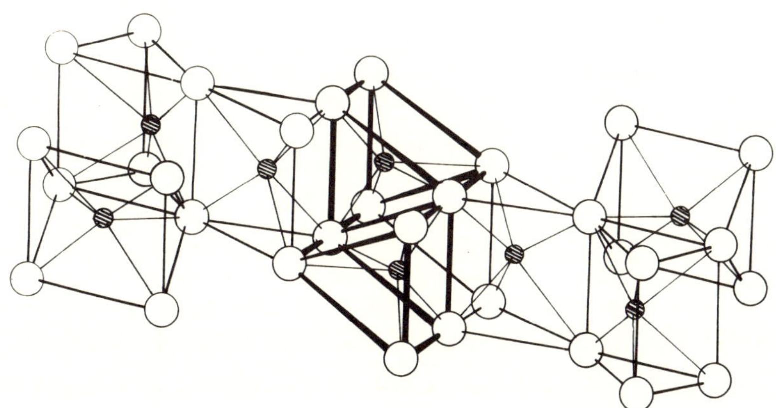

FIGURE 9. The unit cell of the Cr_3C_2 binary phase (Toth 1971). Bold bonds delineate two trigonal prisms condensed as in $[Co_3Ni_7(C-C)(CO)_{15}]^{3-}$ (figure $10a$).

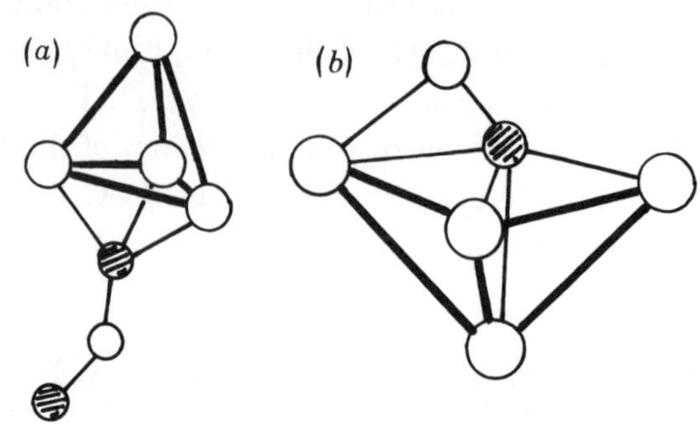

FIGURE 10. Schematic representation of (a) $[Fe_4(CO)_{12}(CO-CH_3)]^-$ (Holt *et al.* 1980; Dawson *et al.* 1980), and (b) $[HFe_4(CO)_{13}]^-$ (Manassero *et al.* 1976).

The high-yield synthesis of cobalt and iron carbide carbonyl clusters here reported confirms that intermolecular dual coordination of a carbonyl group bonded to a molecular cluster with a Lewis acid, such as acetyl or benzoyl carbocation, precedes a ready CO cleavage. In the described approach, CO splitting requires a transfer of electrons from an external source, and is probably favoured by the increase in metal coordination of the carbide atom and by delocalization of the C–O bond in the leaving carboxylate group.

Dual coordinations, related to those represented in figure 10, may also occur on metal crystallites either by interaction of the oxygen atom of a metal-bound carbonyl group with an acidic site of the support (Shriver 1981) or, directly, onto the crystallite metal surface, as well as at surface defects such as steps or adatoms (Chini 1979). The relevance of surface carbide atoms in methanation or Fischer–Tropsch reactions has been recently confirmed by isotopic studies (Sachtler *et al.* 1978).

REFERENCES

Albano, V. G., Braga, D., Ciani, G. & Martinengo, S. 1981 *J. organometall. Chem.* **213**, 293–301.

Albano, V. G., Chini, P., Ciani, G., Sansoni, M. & Martinengo, S. 1980 *J. chem. Soc. Dalton Trans.*, pp. 163–166.

Albano, V. G., Chini, P., Ciani, G., Sansoni, M., Strumolo, D., Heaton, B. T. & Martinengo, S. 1976 *J. Am. chem. Soc.* **98**, 5027–5028.

Albano, V. G., Chini, P., Martinengo, S., Sansoni, M. & Strumolo, D. 1974 *J. chem. Soc. chem. Commun.*, pp. 300–301.

Albano, V. G., Chini, P., Martinengo, S., Sansoni, M. & Strumolo, D. 1978 *J. chem. Soc. Dalton Trans.*, pp. 459–463.

Chini, P. 1979 *Gazz. chim. ital.* **109**, 225–240.

Chini, P. 1980 *J. organometall. Chem.* **200**, 37–61.

Chini, P., Longoni, G. & Albano, V. G. 1974 *Adv. organometall. Chem.* **14**, 285–344.

Chini, P., Martinengo, S. & Garlaschelli, L. 1972 *J. chem. Soc. Chem. Commun.*, pp. 709–710.

Chini, P., Martinengo, S., Strumolo, D., Albano, V. G. & Sansoni, M. 1973 In *Abstr. 165th Nat. Meeting of Am. Chem. Soc., Dallas*, p. 57.

Churchill, M. R., Wormald, J., Knight, J. & Mays, M. J. 1971 *J. Am. chem. Soc.* **93**, 3073–3074.

Colton, R., Commons, C. J. & Hoskins, B. F. 1975 *J. chem. Soc. chem. Commun.*, pp. 363–364.

Dawson, P. E., Johnson, B. F. G., Lewis, J. & Raithby, P. R. 1980 *J. chem. Soc. chem. Commun.*, pp. 781–783.

Hamilton, D. M., Willis, W. S. & Stucky, G. D. 1981 *J. Am. chem. Soc.* **103**, 4255–4256.

Herrman, W. A., Biersack, H., Ziegler, M., Weidehammer, K., Siegel, R. & Rehder, D. 1981 *J. Am. chem. Soc.* **103**, 1692–1699.

Hodali, H. A. & Shriver, D. F. 1979 *Inorg. Chem.* **18**, 1236–1241.

Holt, E. M., Whitmire, K. & Shriver, D. F. 1980 *J. chem. Soc. chem. Commun.*, pp. 778–779.

Holt, E. M., Whitmire, K. & Shriver, D. F. 1981 *J. organometall. Chem.* **213**, 125–137.

Hsieh, A. T. T. & Mays, M. J. 1972 *J. organometall. Chem.* **37**, C53–C55.

Ingle, W. M., Preti, G. & MacDiarmid, A. 1973 *J. chem. Soc. chem. Commun.*, pp. 497–498.

Manassero, M., Sansoni, M. & Longoni, G. 1976 *J. chem. Soc. chem. Commun.*, pp. 919–920.

Nakagura, S. 1957 *J. phys. Soc. Japan* **12**, 482.

Pino, P., Piacenti, F. & Bianchi, M. 1977 In *Organic chemistry via metal carbonyls* (ed. P. Pino & I. Wender), vol. 2, pp. 43–232. New York: Wiley.
Sachtler, J. W. A., Kool, J. M. & Ponec, V. 1979 *J. Catal.* **56**, 284–286.
Shriver, D. F. 1981 In *Catalytic activation of carbon monoxide* (Am. Chem. Soc. Symp. no. 152) (ed. P. C. Ford), pp. 1–18.
Tachikawa, M. & Muetterties, E. L. 1980 *J. Am. chem. Soc.* **102**, 4541–4542.
Tachikawa, M. & Muetterties, E. L. 1982 *Prog. inorg. Chem.* (In the press.)
Toth, L. E. 1971 *Transition metal carbides and nitrides*, pp. 51–52. New York: Academic Press.
Whitmire, K. & Shriver, D. F. 1981 *J. Am. chem. Soc.* **103**, 6754–6755.

Phil. Trans. R. Soc. Lond. A **308**, 59–66 (1982)   [59]
Printed in Great Britain

Nucleophilic addition at μ_3-alkyne clusters leading to carbon–carbon bond formation

By A. J. Deeming and P. J. Manning

*Department of Chemistry, University College London, 20 Gordon Street,
London WC1H 0AJ, U.K.*

Reactions of nucleophiles with triosmium carbonyl clusters, especially those containing unsaturated hydrocarbon ligands, are discussed. Attack may be at CO, the metal atoms, at carbon of the organic ligand, or, where there are acidic metal-bound hydrogen atoms, deprotonation to give anionic clusters may occur.

New results on the reactions of $LiBHEt_3$ with μ_3-alkyne clusters of type $Os_3(CO)_{10}$ (RC_2R') are considered in the light of the range of possible sites of attack. Protonation of anionic species that are formed gives hydrogenation products with or without the loss of CO. $Os_3H_2(CO)_9(RC_2R')$ is usually a minor product, while C–C coupling leads to $Os_3H(CO)_9(CRCR'COH)$ (in general the major product) and to $Os_3H(CO)_9$-($CRCR'CH$). With terminal alkynes RC_2H H-atom transfer accompanies C–C coupling to give $Os_3H(CO)_9(RC{=}C{=}CH_2)$ in substantial amounts. The initial site of hydride attack (CO, alkyne or metal) is considered in the context of low-temperature ^{1}H n.m.r. results.

Introduction

Clusters have modes of reactivity that are unavailable to mononuclear compounds and in addition to those found for mononuclear species. Reaction steps may involve a modification of the metal–metal bonds; changes in order of these bonds or in the geometry of the metal framework are found. Changes in nuclearity are a direct extension of this. Furthermore, certain reaction types are expected to relate to the presence and unique character of bridging ligands. There have been few mechanistic studies on known reactions of clusters, while for the most part we are still at the stage of defining reactions of clusters. Odd-electron processes (radical and redox reactions) may prove in time to be of great importance in cluster chemistry, but most reactions so far identified may be thought of as even-electron processes involving electrophilic and nucleophilic centres, either or both of which may be at the cluster.

It is remarkable how much of our present knowledge of clusters is based on the very simple clusters, $M_3(CO)_{12}$ (M = Fe, Ru, or Os) and trinuclear derivatives of these (see, for example, Johnson 1980). Triosmium compounds are particularly inert and tractable, and their reactions most readily studied. We shall concentrate on these exclusively.

Most carbonyl triosmium clusters are attacked by electrophiles (H$^+$ in particular) at the metal atoms to give bridging metal hydrides or sometimes at the oxygen atoms of CO ligands (Gavens & Mays 1978). Perhaps attack at oxygen precedes subsequent intramolecular transfer of the electrophile (H$^+$) to the metal atoms even in cases where the final product is the metal hydride. Protonation at a carbon atom of organic ligands is rare. One case we have observed recently is the protonation of the vinyloxy-ligand in $Os_3H(CO)_{10}(\mu\text{-OCH}{=}CH_2)$ by trifluoroacetic acid in $CDCl_3$ at $-50\,^\circ C$ to give the ethanal compound $[Os_3H(CO)_{10}(CH_3CHO)][CF_3CO_2]$, which readily converts to $Os_3H(CO)_{10}(CF_3CO_2)$ with the liberation of free ethanal on warming to room temperature (A. J. Arce, A. J. Deeming & J. Watson-Miller, unpublished 1982).

We shall begin with a survey of the attack of nucleophiles at trimetallic clusters, giving examples in the main from our own recent work. The second part of this paper is a description of the addition of hydride ions (as $LiBHEt_3$) to alkyne complexes of the type $Os_3(CO)_{10}$(alkyne) leading to C–C bond formation at the alkynes and other reactions.

Nucleophilic addition at triosmium clusters

(a) Attack at CO

This occurs commonly where there are no organic ligands present that are more electrophilic than CO. $Os_3(CO)_{12}$ reacts with amines RNH_2 to give $Os_3H(CO)_{10}(RNHCO)$ (1), containing the amido ligand bridging two osmium atoms; the reaction occurs readily at room temperature if the neat amine is used (Kaesz *et al.* 1982; Azam *et al.* 1978; Arce & Deeming

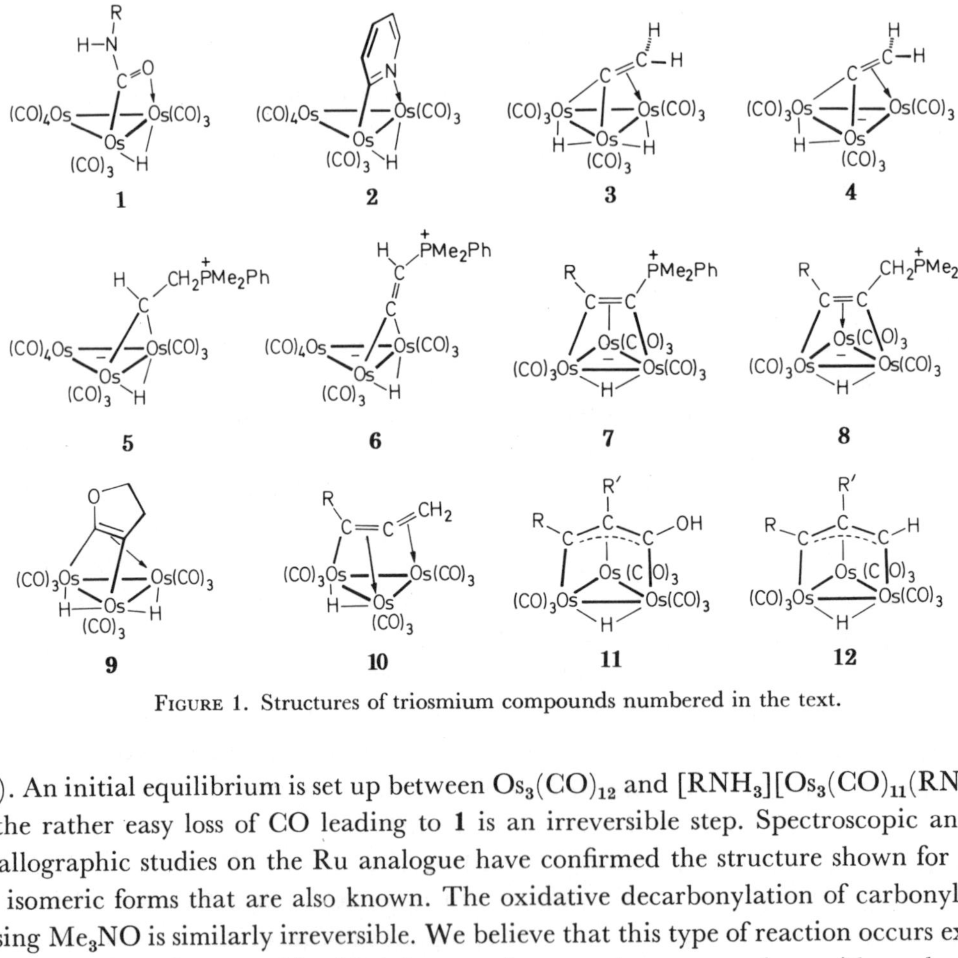

Figure 1. Structures of triosmium compounds numbered in the text.

1980). An initial equilibrium is set up between $Os_3(CO)_{12}$ and $[RNH_3][Os_3(CO)_{11}(RNHCO)]$, but the rather easy loss of CO leading to **1** is an irreversible step. Spectroscopic and X-ray crystallographic studies on the Ru analogue have confirmed the structure shown for **1** rather than isomeric forms that are also known. The oxidative decarbonylation of carbonyl clusters by using Me_3NO is similarly irreversible. We believe that this type of reaction occurs extremely generally and may be an unidentified first step in many known reactions with nucleophiles or is not normally observed because of ready reversibility. For example, the 2-pyridyl compound $Os_3H(CO)_{10}(NC_5H_4)$ (**2**) (Choo Yin & Deeming 1975) dissolves in undiluted benzylamine to give a new hydrido-species in solution with $\nu(CO)$ absorptions just over 30 cm^{-1} lower than for **2**. Addition of small amounts of chloroform to this solution regenerates **2** totally. Corresponding changes are observed in the ^{1}H n.m.r. spectrum and we believe that the species formed is $[RNH_3][Os_3H(CO)_9(NC_5H_4)(RNHCO)]$ (A.J. Arce, A.J. Deeming & R. Shaunak, unpublished 1982). We do not know why the anionic RNHCO species does not lose CO with irreversible product formation in this case, while the loss of CO occurs so readily when $Os_3(CO)_{12}$ or $Ru_3(CO)_{12}$ are dissolved in the undiluted amine.

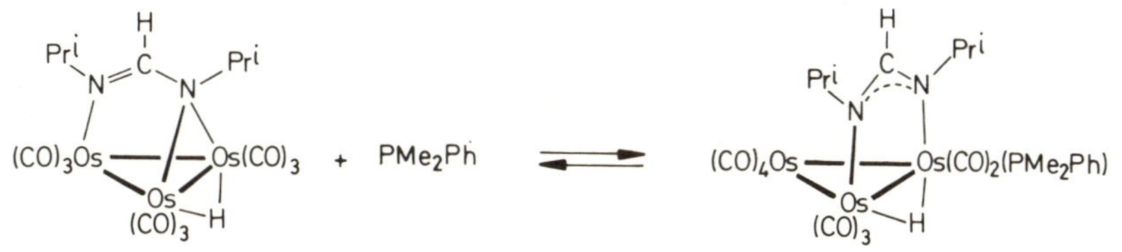

FIGURE 2. Reversible uptake of PMe_2Ph facilitated by a change in the bonding of the $Pr^iNCHNPr^i$ ligand to the triosmium cluster.

(b) Attack at metal

Although direct nucleophilic attack at metal atoms in normal oxidation states is typical behaviour, metal compounds in a low oxidation state usually require special features for such reaction. Essentially all the reactions of $Os_3H_2(CO)_{10}$ may be regarded as going via a direct nucleophilic addition at metal (Shapley *et al.* 1975; Deeming & Hasso 1975a, 1976a). For example, it has been shown recently that many anions, including halide, add to this dihydride directly (Kennedy *et al.* 1981). The compound may not contain a double metal–metal bond but certainly reacts as an unsaturated molecule. Another formally unsaturated species that undergoes additions is $Os_3(CO)_9(PhC_2Ph)$ (Clauss *et al.* 1981), while clusters with weakly bound ligands L such as $Os_3(CO)_{10}L_2$ (L = acetonitrile, for example) presumably undergo reversible loss of L to give reactive unsaturated species. In other cases direct addition at the metal is possible because a ligand already present transforms from an n to an $n-2$ electron donor. In known examples it is not clear whether the modification occurs unimolecularly before nucleophilic addition at the metal or whether the nucleophilic attack induces the transformation. Figure 2 illustrates an example of this behaviour (Deeming & Peters 1982), and similar interconversions also occur with the corresponding compound from 2-aminopyridine (Deeming *et al.* 1982).

Modification of the metal skeleton of the cluster to allow direct nucleophilic attack, although uncommon for triosmium species, becomes a feature of higher clusters. There are a few cases of the transfer of a nucleophile from an organic ligand to the metal. The alkynyl cluster $Os_3H(CO)_9(C{\equiv}CCPh_2OH)$ undergoes an acid-catalysed isomerization to $Os_3H(OH)(CO)_9$-$(C{=}C{=}CPh_2)$ and it is likely that the unsaturated intermediate $[Os_3H(CO)_9(C{=}C{=}CPh_2)]^+$ is attacked at the metal directly by OH^- or H_2O. Nucleophilic attack leads to an opening of one edge of the metal triangle (Aime *et al.* 1982).

(c) Deprotonation

Many triosmium carbonyl clusters contain metal-bound hydrogen atoms that generally show protic rather than hydridic character. Thus nucleophiles, especially good bases, often deprotonate clusters. The vinylidene complex $Os_3H_2(CO)_9(C{=}CH_2)$ (**3**) is readily deprotonated to give the anion (4), which may be isolated as $[N(PPh_3)_2][Os_3H(CO)_9(C{=}CH_2)]$. The equivalence of the vinylidene hydrogen atoms shows that deprotonation generates a plane of symmetry. Deprotonation occurs in spite of the vinylidene ligand's being a potential electrophilic centre even with cyanide ion, which is a good nucleophile towards carbon in most cases.

(d) Attack at organic ligands

Nucleophilic attack commonly occurs at η-alkene (monoene or polyene) in cationic complexes, not so much because of electron withdrawal from the alkene by the metal but because of transition state stabilization caused by a formation of a σ-metal–carbon bond (in the monoene) synchronous with the formation of a bond from carbon to the nucleophile. The ability of the alkene to slide across the metal is important in this respect (Eisenstein & Hoffmann 1981). Neutral metal clusters are attacked, as in the addition of PMe_2Ph to $Os_3H(CO)_{10}(CH{=}CH_2)$ to give 5 (Deeming & Hasso 1976b; Churchill $et\ al.$ 1976). As well as PMe_2Ph, other nucleophiles such as RO^-, $NHEt_2$, ArS^-, and CN^- also add. The alkyne cluster $Os_3(CO)_{10}(C_2H_2)$ gives the PMe_2Ph adduct (6), $Os_3H(CO)_9(C{\equiv}CR)$ gives 7, and $Os_3H(CO)_9(CR{=}C{=}CH_2)$ gives 8. The X-ray structures of 6, 7 (R = H) and 8 (R = Me) have been determined (Henrick $et\ al.$ 1982). In contrast, other alkyne clusters of the type $Os_3(CO)_{10}$(alkyne) do not add PMe_2Ph, other than by slow reaction to give substitution compounds. Only with C_2H_2 can PMe_2Ph attack at a CH terminus, allowing a hydrogen atom to be transferred then from the other carbon of the alkyne. Presumably the initial attack without a subsequent H-atom transfer is unfavourable. The allyl complexes $Os_3H(CO)_9(CRCR'CR'')$ are also unreactive towards PMe_2Ph at room temperature.

Of the examples we have studied, the α-carbon of the alkynyl complexes $Os_3H(CO)_9(C{\equiv}CR)$ appears the most electrophilic, and c.n.d.o. calculations indicate that this carbon carries more positive charge than the adjacent one but less than the carbonyl carbon atoms (Granozzi $et\ al.$ 1982). The hydroxy-alkyne $HC{\equiv}CCH_2CH_2OH$ reacts with $Os_3(CO)_{12}$ to give the dihydrofuran derivative, 9. The same compound is obtained from $Os_3(CO)_{10}(MeCN)_2$ via $Os_3(CO)_{10}$-$(HC_2CH_2CH_2OH)$ and $Os_3H(CO)_9(C{\equiv}CCH_2CH_2OH)$. The latter compound undergoes an intramolecular nucleophilic attack at the α-carbon atom to give 9 (S. Aime & A. J. Deeming, unpublished 1982).

Hydride ion attack at $Os_3(CO)_{10}$(alkyne)

Of the complexes of the type $Os_3(CO)_{10}$(alkyne) only the C_2H_2 complex is readily attacked by PMe_2Ph (see above). The monosubstituted and disubstituted alkyne compounds show no rapid reaction with PMe_2Ph to give zwitterionic adducts, only slowly giving $Os_3(CO)_{10}$-$(PMe_2Ph)_2$. Hydride ions (as $LiBHEt_3$) react rapidly with all the compounds of this type that we have studied, to give, after acidification and chromatographic separation, various products (10, 11 and 12) (figure 1) and $Os_3H_2(CO)_9$(alkyne) (13). Yields and spectroscopic data are given in table 1. A few of these clusters are new, but all are of known types and were easily characterized by comparing their spectroscopic data with those reported. Notably, except for 13, which is formed in low yield in most reactions, all these compounds result from the reduction of a CO ligand and its coupling to the alkyne ligand. The hydroxyallyl compounds (11) were previously formed by reacting $Os_3(CO)_{11}(MeCN)$ with the alkyne RC_2H (R = H, Me or Ph) in moist CH_2Cl_2 but not with disubstituted alkynes (Hanson $et\ al.$ 1980). By our route, compounds 11 were synthesized from but-2-yne and diphenylacetylene (R = R' = Me or Ph). The

TABLE 1. YIELDS AND SPECTROSCOPIC DATA FOR PRODUCTS FROM TREATING $Os_3(CO)_{10}$(ALKYNE) WITH $LiBHEt_3$

alkyne	products	yield (%)	$\nu(CO)/cm^{-1}$[†]					selected ^1H n.m.r. data; δ[‡]
HC_2H	$Os_3H^w(CO)_9(CH^x{=}C{=}CH_2{}^{y,\,z})$	2	2101m 2003s	2072vs 1994m	2047vs 1986w	2026vs 1982mw	2017s	9.11d (H^x) (J = 1.9 Hz), 4.30s, 2.61s (H^yH^z), − 22.43d (H^w)
	$Os_3H(CO)_9(CHCHCOH)$	68§						
	$Os_3H(CO)_9(CHCHCH)$	3	2100 2004s	2072s 1990w	2051s 1979w	2021s	2010m	
	$Os_3H_2(CO)_9(HC_2H)$	5	2110m 2013s	2082s 2004m	2058s 1991sh	2033s 1986m	2026m	8.63d (C_2H_2) (J = 1.3 Hz), − 21.62††, − 18.50d (J = 1.0 Hz)
MeC_2H	$Os_3H(CO)_9(MeC{=}C{=}CH_2)$	43	‖					
	$Os_3H^x(CO)_9(CH^yCMeCOH)$¶	16	2098m 2046sh 2001sh	2096sh 2022vs 1998vs	2068vs 2018sh 1995sh	2066sh 2012m 1990sh	2049vs 2007s	2.89s (Me), 7.67d (H^y) (J = 1.5 Hz), − 18.80d (H^x)
	$Os_3H(CO)_9(MeCCHCH)$	9	‖					
	$Os_3H_2(CO)_9(MeC_2H)$	7	2110m 2012vs	2081vs 2003s	2058vs 1985s	2033vs	2026s	8.03s (CH), 2.81s (Me), − 17.97s, − 21.49s
PhC_2H	$Os_3H^x(CO)_9(PhC{=}C{=}CH_2{}^{y,\,z})$	25	2100s 2003s	2072vs 1993m	2048vs 1987w	2026vs 1982m	2019s	4.14d (H^y), 2.65d (H^z) (J = 2 Hz), − 22.34s (H^x)
MeC_2Me	$Os_3H(CO)_9(CMeCMeCOH)$	22	2095m 2003s	2068sh 1996vs	2065vs 1986sh	2046vs 1957vw	2014vs	2.12s (Me), 2.93s (Me), − 18.69s (OsH)
	$Os_3H^x(CO)_9(CMeCMeCH^y)$	7	2099w 2002s	2070vs 1990w	2048vs	2020s	2008m	2.86s (Me), 2.51s (Me), 8.78s (H^y), − 19.23s (H^x)
	$Os_3H_2(CO)_9(MeC_2Me)$	26	2106w 2009vs	2080vs 1997s	2054vs 1981m	2030vs	2020m	2.49s (Me_2), − 17.99s (OsH), − 21.46s (OsH)‡‡
PhC_2Ph	$Os_3H^x(CO)_9(CPhCPhCOH^y)$	51	2099s 2005vs	2074vs 1988sh	2043vs 1983m	2024vs	2011s	6.4s (H^y) (temp. variable), − 18.78s (H^x)
	$Os_3H_2(CO)_9(PhC_2Ph)$ ‖	trace	2108m 2015vs	2083vs 1999s	2057vs 1995sh	2032vs 1983s	2027vs	− 17.30s, − 21.21s
	$Os_3H(CO)_9(CPhCPhCH)$	trace						

† In cyclohexane solution.   ‡ In $CDCl_3$.   § Hanson *et al.* (1980).   ‖ Deeming *et al.* (1975).   ¶ Isomeric $Os_3H(CO)_9(CMeCHCOH)$ (Hanson *et al.* 1980).   †† Apparent quartet; doublet of triplets with equal coupling to the C_2H_2 protons, − 17 °C.   ‡‡ Coalesced hydride signal at − 19.43; T_c 1 °C.

complex from propyne has distinctly different ^{1}H n.m.r. data from those reported for $Os_3H(CO)_9(MeCCHCOH)$, so our compound may have the opposite regiospecificity, that is $Os_3H(CO)_9(CHCMeCOH)$. These compounds exist as only slowly interconverting isomers, as already described (Hanson *et al.* 1980).

Other products, **10** and **12**, involve further reduction of the coupled CO with the loss of an oxygen atom. These compounds seem to be favoured when rather larger excesses of $LiBHEt_3$ (such as a fivefold excess) are used.

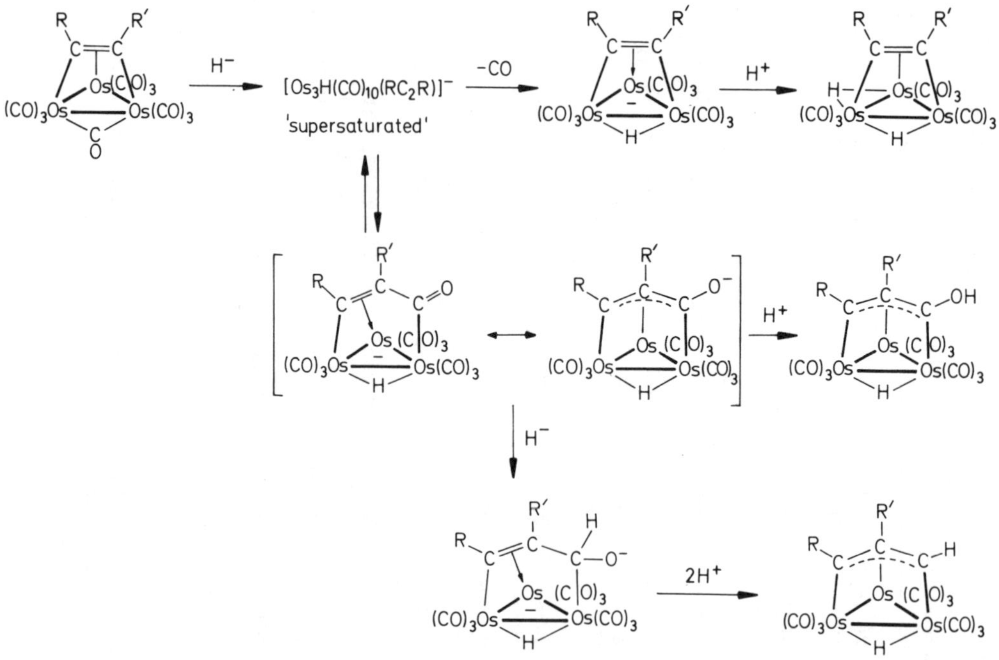

FIGURE 3. A possible reaction pathway resulting from the treatment of $OS_3(CO)_{10}$(alkyne) with $LiBHEt_3$ involving hydride addition at metal atoms. The compound $[Os_3H(CO)_{10}$(alkyne)]$^-$ is only supersaturated if the alkyne remains as a μ_3-four-electron donor and if there is no Os–Os bond cleavage.

To learn of the origin of compounds **10–13** we studied the addition of $LiBHEt_3$ (twofold excess) in THF to $Os_3(CO)_{10}(C_2H_2)$ in d^8-THF at -20 °C. Rapid changes in the ^{1}H n.m.r. spectrum were observed. The single C_2H_2 resonance is replaced by ten new signals between $\delta = 6$ and 15, together with two hydride signals. Clearly the system is complicated. One set of signals observed immediately ($\delta = 7.90$d, 6.65d and -18.10s) corresponds to the deprotonated form of the hydroxyallyl, $[Os_3H(CO)_9(CHCHCO)]^-$, and this persists indefinitely at this temperature. The deprotonated form of $Os_3H_2(CO)_9(C_2H_2)$ is only slowly formed at the expense of other signals in the spectrum (spectrum of $[Os_3H(CO)_9(C_2H_2)]^-$, $\delta = 8.60$s and -21.13s). After several hours at -20 °C the spectrum had simplified and the solution contained mainly $[Os_3H(CO)_9(C_2H_2)]^-$ and $[Os_3H(CO)_9(CHCHCO)]^-$, and protonation with CF_3CO_2H gave $Os_3H_2(CO)_9(C_2H_2)$ and $Os_3H(CO)_9(CHCHCOH)$. These were each isolated in approximately 30% yield.

Very weak signals ($\delta = 14.9$ and 13.6) in the early stages of the reaction could be of formyl species but after several hours these had disappeared. The room temperature addition of $LiBHEt_3$ (4 mol per Os_3) also gave $Os_3H(CO)_9(CHCHCH)$ and $Os_3H(CO)_9(CH{=}C{=}CH_2)$ in low yield, but these were not observed in the reaction at -20 °C. Presumably they are formed in secondary reactions.

Two basic but not necessarily exclusive reaction schemes are shown in figures 3 and 4. The route via formyl intermediates in figure 4 is supported by the observation of weak low-field signals, but we believe that the route would not result in C–C coupling until after protonation. The formation of the CHOH and CH_2 complexes is based on the route to $Os_3(CO)_{11}(CH_2)$ by hydride ion attack at $Os_3(CO)_{12}$ (Steinmetz & Geoffroy 1981). The final stages of the scheme in figure 4 require isomerizations with alkyne to CH_2 or CHOH coupling. There is some precedent for this in the isomerization of $Os_3(CO)_9(CH_2)(PhC_2Ph)$ to $Os_3H(CO)_9(CPhCPhCH)$, but this requires a temperature of 135 °C and does not occur at room temperature (Clauss *et al.* 1981). We favour the scheme in figure 3 as the main route. Addition of hydride at the

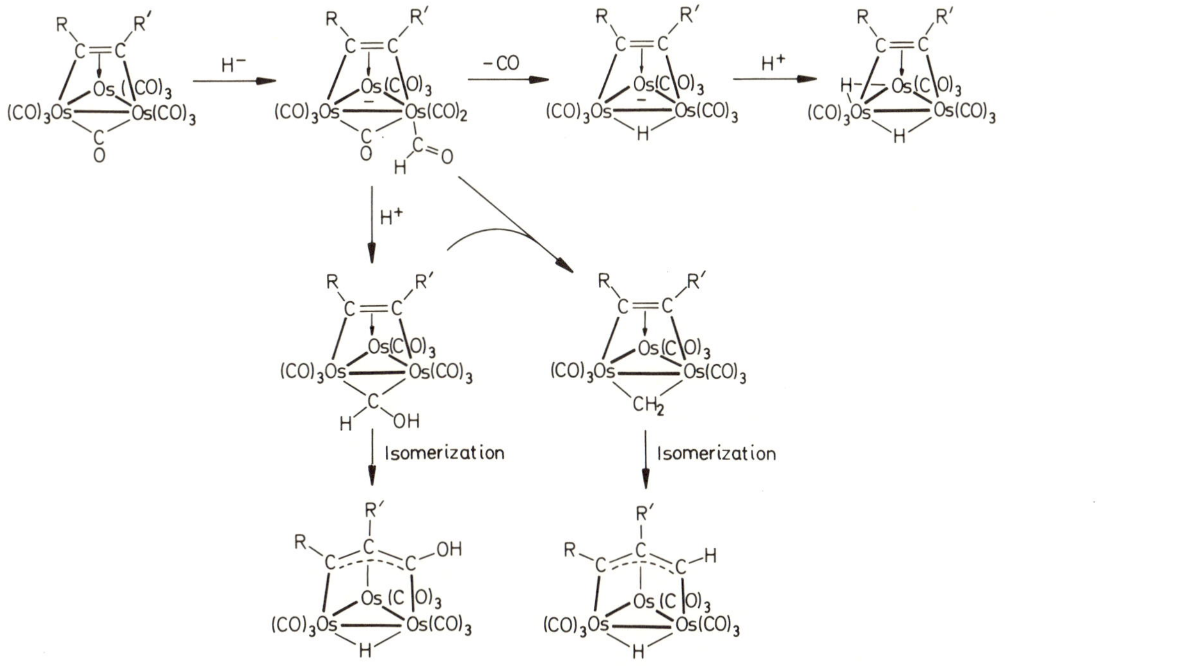

FIGURE 4. An alternative reaction pathway based on hydride addition at CO to give intermediate formyl species.

metal atoms, or possibly an initial attack at C_2H_2 or CO followed by H-atom transfer to the metal atoms, would give $[Os_3H(CO)_{10}(C_2H_2)]^-$. This is supersaturated if the alkyne remains a μ_3-four-electron-donor ligand. This could then lose CO or undergo C–C coupling. The latter might be favoured by a four-electron to two-electron donor transformation of the alkyne. Interestingly C_2H_2 reacts with $Os_3(CO)_{11}(CH_3CN)$ with C–C coupling and the presumed intermediate, $Os_3(CO)_{11}(C_2H_2)$ is closely related in electron-counting to $[Os_3H(CO)_{10}(C_2H_2)]^-$. Coupling of CO with alkyne is well known, of course, but not studied mechanistically. The coupling may be rapidly reversible in some cases at least (Dyke *et al.* 1980; Finnimore *et al.* 1980) and the slow formation of $[Os_3H(CO)_9(C_2H_2)]^-$ by CO loss may result from a reversal of C–C coupling.

Experimental

We synthesized the alkyne complexes $Os_3(CO)_{10}$(alkyne) by treating $Os_3(CO)_{10}(MeCN)_2$ with the alkyne. All these alkyne complexes have the structure shown in figures 3 and 4 with bridging carbonyl except for the PhC_2Ph complex, which has only terminal CO in the crystal. In solution, weak absorptions due to bridging CO are observed even for $Os_3(CO)_{10}(PhC_2Ph)$. Between 60 and 270 mg of the alkyne complexes were used for the hydride addition. Yields in table 1 are given for the isolated compounds and the reaction conditions were typically as given in the example below.

Reaction of $Os_3(CO)_{10}(C_2H_2)$ with hydride

On treating a solution of $Os_3(CO)_{10}(C_2H_2)$ (0.110 g) in freshly distilled dry THF (30 cm^3) with a fourfold excess of $LiBHEt_3$ in THF (0.5 cm^3) at room temperature, the colour of the solution changed from bright orange to yellow. Trifluoroacetic acid (0.050 cm^3) was added after 8 h. After 20 h the solvent was removed under reduced pressure and chromatographic separation of the residue (SiO_2, t.l.c.) eluting with petroleum ether gave bands that yielded $Os_3H_2(CO)_9(C_2H_2)$ (0.005 g, 5%), $Os_3H(CO)_9(CHCHCH)$ (0.003 g, 3%), and $Os_3H(CO)_9$-$(CH{=}C{=}CH_2)$ (0.002 g, 2%). Most of the material remaining on the baseline was eluted with a petroleum ether–chloroform–diethyl ether mixture (80:15:5 by volume) to give two bands. One remains unidentified while the other gave $Os_3H(CO)_9(CHCHCOH)$ (0.074 g, 68%) as pale yellow crystals. Addition of $LiBHEt_3$ at $-78\ ^\circ C$ followed by warming to room temperature did not materially alter the reaction products. The compounds $Os_3H(CO)_9$-$(CH{=}C{=}CH_2)$ and $Os_3H(CO)_9(CHCHCH)$ appear identical to those formed by photolysis then thermolysis of $Os_3(CO)_{11}(CH_2CCH_2)$ which is formed itself from allene and $Os_3(CO)_{11}(MeCN)$ (Johnson et al. 1982).

We thank the S.E.R.C. for a research studentship and Johnson–Matthey Ltd for a generous gift of osmium tetraoxide.

References

Aime, S., Deeming, A. J., Hurtshouse, M. B. & Backer-Dirks, J. D. J. 1982 J. chem. Soc. Dalton Trans., pp. 1625–1629.

Arce, A. J. & Deeming, A. J. 1980 J. chem. Soc. chem. Commun., pp. 1102–1103.

Azam, K. A., Choo Yin, C. & Deeming, A. J. 1978 J. chem. Soc. Dalton Trans., pp. 1201–1206.

Choo Yin, C. & Deeming, A. J. 1975 J. chem. Soc. Dalton Trans., pp. 2091–2096.

Churchill, M. R., DeBoer, B. G., Shapley, J. R. & Keister, J. B. 1976 J. Am. chem. Soc. 98, 2357–2358.

Clauss, A. D., Shapley, J. R. & Wilson, S. R. 1981 J. Am. chem. Soc. 103, 7387–7388.

Deeming, A. J. & Hasso, S. 1975a J. organometall. Chem. 88, C21–C23.

Deeming, A. J. & Hasso, S. 1976a J. organometall. Chem. 114, 313–324.

Deeming, A. J. & Hasso, S. 1976b J. organometall. Chem. 113, C39–C42.

Deeming, A. J., Hasso, S. & Underhill, M. 1975b J. chem. Soc. Dalton Trans., pp. 1614–1620.

Deeming, A. J. & Peters, R. 1982 J. organometall. Chem. (In the press.)

Deeming, A. J., Peters, R., Hurtshouse, M. B. & Backer-Dirks, J. D. J. 1982 J. chem. Soc. Dalton Trans., pp. 1205–1211.

Dyke, A. F., Knox, S. A. R., Naish, P. J. & Taylor, G. E. 1980 J. chem. Soc. chem. Commun., pp. 409–410.

Eisenstein, O. & Hoffmann, R. 1981 J. Am. chem. Soc. 103, 4308–4320.

Finnimore, S. P., Knox, S. A. R. & Taylor, G. E. 1980 J. chem. Soc. chem. Commun., pp. 411–412.

Gavens, P. D. & Mays, M. J. 1978 J. organometall. Chem. 162, 389–401.

Granozzi, G., Tondello, E., Bertoncello, R., Aime, S. & Osella, D. 1982 (Submitted.)

Hanson, B. E., Johnson, B. F. G., Lewis, J. & Raithby, P. R. 1980 J. chem. Soc. Dalton Trans., pp. 1852–1857.

Henrick, K., McPartlin, M., Deeming, A. J., Hasso, S. & Manning, P. 1982 J. chem. Soc. Dalton Trans., pp. 899–906.

Johnson, B. F. G. (ed.) 1980 Transition metal clusters. London and New York: Wiley.

Johnson, B. F. G., Lewis, J., Raithby, P. R. & Sankey, S. W. 1982 J. organometall. Chem. 231, C65–C69.

Kaesz, H. D., Knobler, C. B., Andrews, M. A., van Buskirk, G., Szostak, R., Strouse, C. E., Lin, Y. C. & Mayr, A. 1982 Pure appl. Chem. 54, 131–144.

Kennedy, S., Alexander, J. J. & Shore, S. G. 1981 J. organometall. Chem. 219, 385–395.

Shapley, J. R. & Keister, J. B., Churchill, M. R. & DeBoer, B. G. 1975 J. Am. chem. Soc. 97, 4145–4146.

Steinmetz, G. R. & Geoffroy, G. L. 1981 J. Am. chem. Soc. 103, 1278.

Phil. Trans. R. Soc. Lond. A **308**, 67–73 (1982)   [67]
Printed in Great Britain

Carbon–carbon bond formation at dinuclear metal centres

By R. E. Colborn, A. F. Dyke, S. A. R. Knox,
Kirsty A. Macpherson, K. A. Mead, A. G. Orpen,
J. Roué and P. Woodward

*School of Chemistry, University of Bristol, Cantock's Close,
Bristol BS8 1TS, U.K.*

The Fischer–Tropsch synthesis of hydrocarbons probably involves an array of carbon–carbon bond-formation processes occurring to unite carbene, carbyne, alkyl, olefin, and related species on a metal surface. In seeking to understand the nature of such processes, model diruthenium complexes have been prepared and the products of their thermolysis and reactions with unsaturated hydrocarbons investigated. The combination at a diruthenium centre of two carbenes, of a carbene and an alkyne, and of a carbyne and an olefin is described, and the possible implications for metal surface processes are emphasized.

1. Introduction

The transition metal catalysed Fischer–Tropsch synthesis of hydrocarbons by the hydrogenation of carbon monoxide is believed to involve an initial dissociative chemisorption of CO and H_2 on the metal surface. This provides a surface 'carbide', which is rapidly and successively hydrogenated to yield surface CH, CH_2 and CH_3 species. These then combine to produce the characteristic longer-chain hydrocarbons. It is the way (or ways) in which this combination occurs that is obscure. Strong evidence for the crucial role played by surface methylene groups has been obtained by Pettit & Brady (1980), who, in a revival of the original suggestion of Fischer & Tropsch, see the carbon chain growth occurring through alkyl–methylene combination

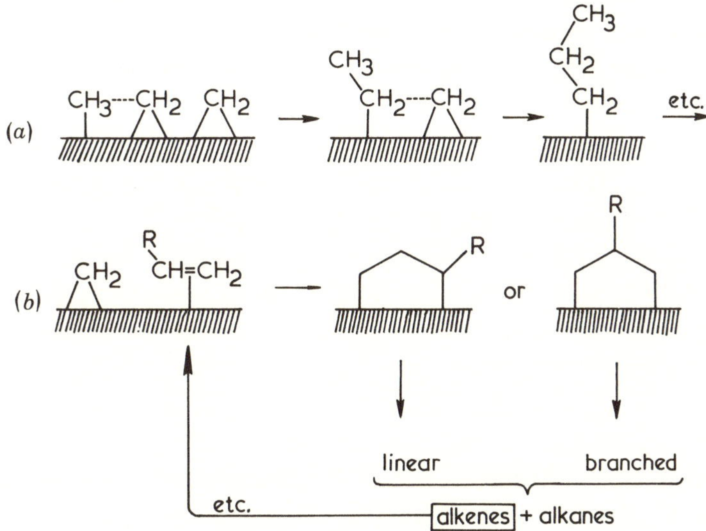

Figure 1. The Pettit (*a*) and Basset (*b*) proposals for carbon chain growth in Fischer–Tropsch synthesis.

on the metal surface (figure 1 a). More recently, Basset *et al.* (1981) have proposed that chain growth could proceed by olefin–methylene combination (figure 1 b). In each case the chain is terminated by β-elimination or reductive elimination with surface hydrogen, or both, generating an olefin or alkane. Whereas, without refinement, the Pettit mechanism accounts only for the production of the high yield of straight-chain hydrocarbons, the Basset mechanism encompasses both straight-chain and branched-chain hydrocarbon formation. The mechanisms should not, however, be seen as mutually exclusive. It is likely that an array of carbon–carbon bond formation processes occurs on the metal surface to unite carbene, carbyne, alkyl, olefin, and related species, leading to the typical wide spectrum of hydrocarbon products of the Fischer–Tropsch synthesis.

In seeking to understand the nature of carbon–carbon bond formation processes at metal centres, model diruthenium complexes have been prepared and their reactivities studied. This paper describes some aspects of these studies.

2. Combination of two carbenes

The di-μ-carbene complexes $[Ru_2(CO)_2(\mu\text{-}CMe_2)(\mu\text{-}CHMe)(\eta\text{-}C_5H_5)_2]$ (**1**) and $[Ru_2(CO)_2(\mu\text{-}CMe_2)(\mu\text{-}CCH_2)(\eta\text{-}C_5H_5)_2]$ (**2**) are readily prepared (Knox *et al.* 1981 a). When the former was treated with ethyne under u.v. irradiation, the expected (Knox *et al.* 1980 b) 'insertion' of the alkyne into a μ-carbene did not occur. Instead, the complex $[Ru_2(\mu\text{-}CO)(\mu\text{-}C_2H_2)(\eta\text{-}C_5H_5)_2]$ (**3**) was formed; both μ-carbenes had been displaced and a four-electron μ-ethyne ligand bridged a ruthenium–ruthenium double bond (see later). The possibility that the carbenes had coupled to produce an olefin led us to study the thermal decompositions of **1** and **2**.

Heating **1** at 200 °C *in vacuo* resulted in the identification, by g.l.c., of the organic products: ethylene (21% of total), propylene (17%), $Me_2C\text{=}C(H)Me$ (46%), $CH_2\text{=}C(Me)Et$ (14%), and $CH_2\text{=}C(H)(^{i}Pr)$ (4%). The formation of ethylene and propylene clearly results from the separate release of :CHMe and :CMe₂ from the diruthenium centre, and subsequent rearrangement. However, the remaining products each contain a five-carbon skeleton derived from the coupling of CHMe with CMe₂. The absence of cross-coupling products with four-carbon or six-carbon skeletons, such as butenes or $Me_2C\text{=}CMe_2$, establishes that the carbene coupling occurs at the diruthenium centre rather than in the gas phase subsequent to release. It may be noted that an X-ray diffraction study of **1** revealed a non-bonding C…C distance of 3.20 Å† separating the two μ-carbenes (Knox *et al.* 1981 a).

Heating **2** to 200 °C resulted in the formation of a hydrocarbon product mixture very similar to that obtained from **1**. Clearly an intramolecular carbene coupling again occurs, but either in association with, or preceded by, hydrogenation of the vinylidene ligand.

The detailed mechanism of combination is not known (e.g. whether a transiently terminal carbene is involved), but an insight has recently been obtained by Shapley *et al.* (1981*b*), who observed that two methylene groups combine at a diosmium centre to provide an $Os_2(\mu\text{-}H)$-$(\mu\text{-}CHCH_2)$ system with bridging hydride and vinyl. Such a system is known to generate ethylene.

The studies described in this section suggest how olefins could be created by combination of carbenes at a dimetal centre on a metal surface. In this way the Basset mechanism of carbon chain growth could be initiated.

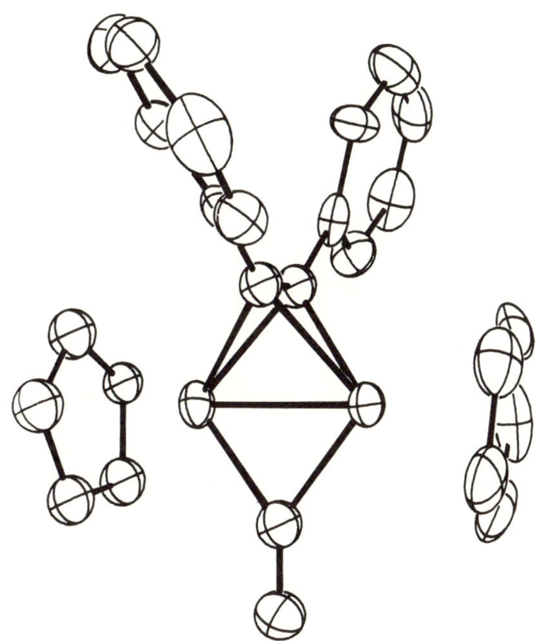

Figure 2. The molecular structure of $[Ru_2(\mu\text{-}CO)(\mu\text{-}C_2Ph_2)(\eta\text{-}C_5H_5)_2]$ (4).

3. Combination of carbene with alkyne

The complex $[Ru_2(\mu\text{-}CO)(\mu\text{-}C_2Ph_2)(\eta\text{-}C_5H_5)_2]$ (4), a diphenylacetylene analogue of 3, has been subjected to X-ray diffraction study. Significant features, seen in figure 2, are a laterally bridging four-electron alkyne ligand and a short metal–metal separation of 2.505(1) Å, according with the ruthenium–ruthenium double bond required by 18-electron rule considerations. In contrast, in complex 1 a formally single Ru–Ru bond of 2.701(1) Å is observed.

Herrmann (1982) has established that an unsaturated dinuclear metal centre, like that in 4, can be expected to react readily with a diazoalkane to generate a μ-carbene system. Complex 4 does indeed react readily, at room temperature, with diazomethane but there is incorporation of two methylene groups rather than the one anticipated. The product is $[Ru_2(CO)$-$(\mu\text{-}CH_2)\{\mu\text{-}C(Ph)C(Ph)CH_2\}(\eta\text{-}C_5H_5)_2]$ (5), whose structure was firmly established by X-ray diffraction and is displayed in figure 3. The CO that was bridging in 4 is now terminally bound, and has been replaced by a bridging methylene group. The other methylene is linked with the alkyne in a C_3 ligand of a type previously produced in reactions of alkynes with μ-carbene complexes (Knox *et al.* 1980*b*). There are grounds for viewing this ligand in the way it is depicted in 5, i.e. as a coordinated vinylcarbene. The presence of two bridging carbon

† $1\ \text{Å} = 10^{-10}\ \text{m} = 10^{-1}\ \text{nm}$.

atoms associated with the diruthenium unit, and previous experience with complexes **1** and **2** (see §2), led us to envisage that thermolysis of **5** would effect their linking, producing perhaps 2,3-diphenylbuta-1,3-diene. Heating **5** to 200 °C did provide six hydrocarbon products; these are not at present identified, but certainly none of them is that diene. A possible explanation of this apparent failure to link the bridging carbons arose in the molecular structure of **5**.

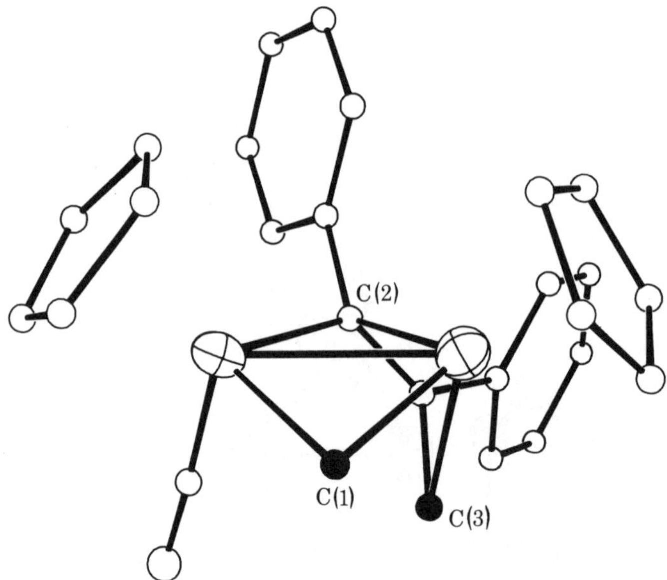

FIGURE 3. The molecular structure of $[Ru_2(CO)(\mu\text{-}CH_2)\{\mu\text{-}C(Ph)C(Ph)CH_2\}(\eta\text{-}C_5H_5)_2]$ (**5**).

The two carbons under discussion, C(1) and C(2) of figure 3, are 3.07 Å apart, but two other carbons, C(1) and C(3), are much closer, at 2.78 Å. These are the methylene carbons introduced via diazomethane. The possibility that carbon–carbon bond formation occurred preferentially to link C(1) and C(3) of this molecule was confirmed when heating **5** in boiling xylene yielded $[Ru_2(CO)(\mu\text{-}CO)\{\mu\text{-}C(Ph)C(Ph)C(H)Me\}(\eta\text{-}C_5H_5)_2]$ (**6**), a compound available to us by another route. Linking of the two methylene groups, and a subsequent hydrogen shift, is associated with scavenging of CO from the presumed decomposition of **5**. Treating **5** with CO under 10 atm (*ca.* 1330 Pa) pressure at 140 °C did, as expected, afford **6**.

The sequence **4** → **5** → **6** represents a carbon chain growth with features strongly reminiscent of the Basset proposal for the mechanism of Fischer–Tropsch synthesis. The step **5** → **6** can thus be seen as a 'methylene plus olefin' combination, leading to a longer-chain olefin. Future work with derivatives of **5**, generated from reactions of **4** with other diazoalkanes, will attempt to clarify the factors controlling the carbon–carbon bond formation. The essential feature of this section is that the μ-methylene carbon in **5** becomes linked not to the other μ-carbon but to the carbon of the C_3 ligand to which it is closest.

In a recent report, Shapley *et al.* (1981*a*) has described the thermally induced linking of μ-methylene and μ-alkyne ligands separated by 2.86 Å on a triosmium cluster.

4. Combination of olefin with carbyne

We (Knox *et al.* 1980*a*) have reported the synthesis of the μ-methylcarbyne complex $[Ru_2(CO)_2(\mu\text{-}CO)(\mu\text{-}CMe)(\eta\text{-}C_5H_5)_2][BF_4]$ (**7**). Under u.v. irradiation in tetrahydrofuran solution the cation reacts with ethylene or propylene to give new complexes in which the olefin and μ-carbyne have become linked. These products $[Ru_2(CO)(\mu\text{-}CO)\{\mu\text{-}C(Me)C(R)CH_2\}(\eta\text{-}C_5H_5)_2]$ (**8**, R = H; **9**, R = Me) are neutral, and deprotonation has evidently occurred,

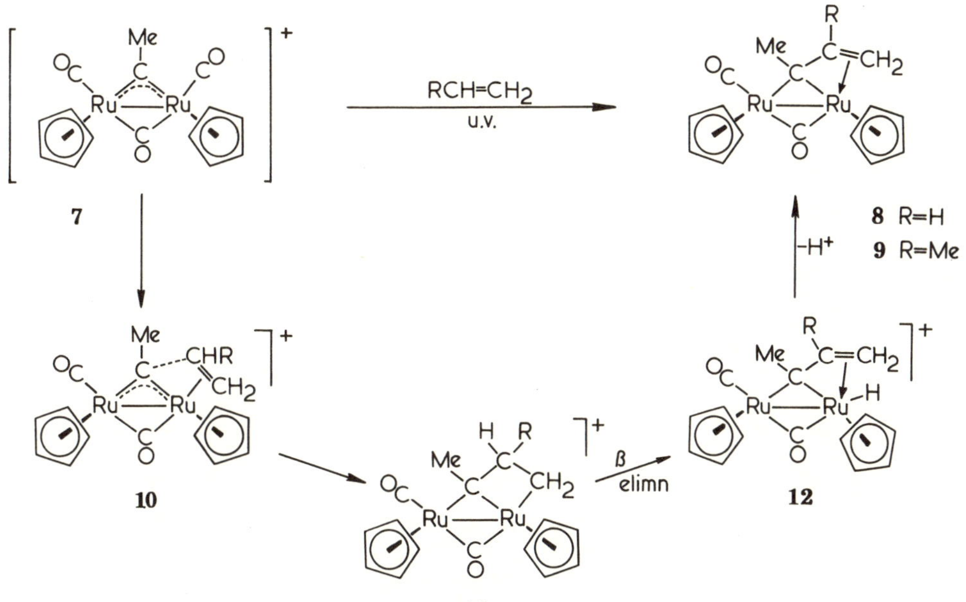

RCH=CH$_2$, u.v.

7

8 R=H **9** R=Me

−H⁺

10

11

12

probably effected by the tetrahydrofuran solvent. An X-ray diffraction study (Knox *et al.* 1981*c*) on **9** confirms the conclusions, reached from n.m.r. spectra, that the new carbon–carbon bond has been formed exclusively between the μ-carbyne carbon and the carbon of propylene that bears the methyl group.

The likely pathway for the formation of **8** and **9** is one in which an initial photochemically induced dissociation of CO from **7** allows the terminal coordination of an olefin. A subsequent intramolecular electrophilic attack of the μ-carbyne carbon upon the olefin would then generate a metallocycle, which through a β-elimination and loss of a proton evolves the observed products. This sequence is shown as **7 → 10 → 11 → 12 → 8/9**. Strong evidence for its authenticity comes from the treatment of the ethylene complex $[Ru_2(CO)(C_2H_4)(\mu\text{-}CO)_2(\eta\text{-}C_5H_5)_2]$ (**13**) (Knox *et al.* 1981*b*) with methyl lithium, tetrafluoroboric acid and sodium borohydride successively, to afford **8**. The first two of these reagents are well known (Knox *et al.* 1981*a*) to convert μ-CO to μ-CMe⁺ at a diruthenium centre, and there can therefore be little doubt of the intermediacy of **10** in this synthesis. The function of $NaBH_4$ would be to aid deprotonation of **12**.

An important feature of the work described in this section is that carbon–carbon bond formation occurs only between the μ-carbyne and the C(Me) carbon of propylene. It is not at present clear why this should be so. Simplistically, it could be argued that an electrophilic attack on a coordinated olefin should occur at the most electron-rich carbon, i.e. that bearing the methyl group, but this does not take into account the stability of the metallacarbonium ion formed, which is difficult to assess. Moreover, steric factors have been previously observed to be important in determining the geometry of such organodirutheniumdicyclopentadienyl complexes, and the methyl group of **9** is seen by X-ray diffraction to occupy a site of low crowding.

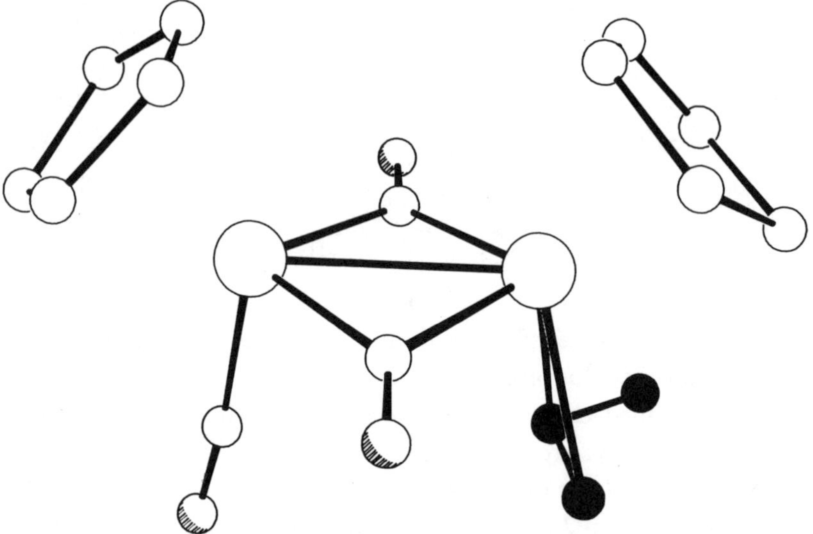

13 R=H

14 R=Me

An X-ray diffraction study on **14** provides no basis for the specificity; the coordination of the propylene is unremarkable and the bound carbons are nearly equidistant from the bridging sites of the diruthenium centre (figure 4).

FIGURE 4. The molecular structure of $[Ru_2(CO)(C_2H_3Me)(\mu\text{-}CO)_2(\eta\text{-}C_5H_5)_2]$ (**14**).

The combination of μ-carbyne and propylene in the way observed leads to a branched carbon skeleton. There is significance here for the Fischer–Tropsch synthesis in that such a linking of carbyne and α-olefin on a metal surface could, after β-elimination or reductive elimination, or both, provide a route to the branched hydrocarbons which are formed in low yield.

5. Conclusions

There are indications that carbon–carbon bond formation processes may occur readily at dinuclear metal centres to link together a wide variety of simple hydrocarbon species, but much study remains before an understanding of the factors controlling these processes can be achieved.

We are grateful to the S.E.R.C. for the award of Research Studentships (A.F.D. and K.A.M.) and for support, to the National Science Foundation (U.S.A.) and Nato for the award of a Fellowship (R.E.C.), and to Johnson Matthey and Co. for a generous gift of ruthenium trichloride.

References

Basset, J.-M., Hugues, F., Besson, B., Bussière, P., Dalmon, J. A. & Olivier, D. 1981 *Nouv. J. Chim.* **5**, 207–210.

Herrmann, W. A. 1982 *Adv. organometall. Chem.* **20**, 159–263.

Knox, S. A. R., Cooke, M., Davies, D. L., Guerchais, J. E., Mead, K. A., Roué, J. & Woodward, P. 1981*a* *J. chem. Soc. chem. Commun.*, pp. 862–864.

Knox, S. A. R., Davies, D. L., Dyke, A. F., Endesfelder, A., Naish, P J., Orpen, A. G., Plaas, D. & Taylor, G. E. 1980*a* *J. organometall. Chem.* **198**, C43–C49.

Knox, S. A. R., Davies, D. L., Dyke, A. F. & Morris, M. J. 1981*b* *J. organometall. Chem.* **215**, C30–C32.

Knox, S. A. R., Dyke, A. F., Guerchais, J. E., Roué, J., Short, R. L., Taylor, G. E. & Woodward, P. 1981*c* *J. chem. Soc. chem. Commun.*, pp. 537–538.

Knox, S. A. R., Dyke, A. F., Naish, P. J. & Taylor, G. E. 1980*b* *J. chem. Soc. chem. Commun.*, pp. 803–805.

Pettit, R. & Brady, R. C. 1980 *J. Am. chem. Soc.* **102**, 6181–6182.

Shapley, J. R., Clauss, A. D. & Wilson, S. R. 1981*a* *J. Am. chem. Soc.* **102**, 6181–6182.

Shapley, J. R., Sievert, A. C., Churchill, M. R. & Wasserman, H. J. 1981*b* *J. Am. chem. Soc.* **103**, 6975–6977.

Phil. Trans. R. Soc. Lond. A **308**, 75–83 (1982)   [75]
Printed in Great Britain

Some theoretical and structural aspects of gold cluster chemistry

By D. M. P. Mingos

Inorganic Chemistry Laboratory, University of Oxford,
*South Parks Road, Oxford OX*1 *3QR, U.K.*

The bonding in tertiary phosphine cluster compounds of gold is sufficiently straight-forward to permit an effective interaction between theoretical concepts developed from semi-empirical molecular orbital calculations and synthetic and structural chemistry. At the simplest conceptual level the isolobal nature of the $Au(PR_3)$ fragment and either the CH_3 or H radicals provides a basis for understanding the structures of a wide range of homonuclear and heteronuclear clusters, e.g. $Os_3(CO)_{10}$-$H(AuPPh_3)$ and $(OC)_5VAu_3(PPh_3)_3$. However, this simplified approach neglects some secondary gold–gold interactions between adjacent gold atoms, which arise from the availability of the higher-lying gold 6p orbitals.

In low-nuclearity clusters tetrahedral fragments, which permit the effective formation of four-centre two electron bonds between the $Au(PR_3)$ fragments, are preferred to larger deltahedra. In higher-nuclearity clusters the stabilities of the clusters depend on the presence of a central gold atom that provides strong radial gold–gold bonding. The relative importance of the radial and tangential components to the total bonding has been effectively demonstrated by a structural comparison of alternative $Au_9(PR_3)_8^{3+}$ clusters. The predictive capability of the theoretical approach has been demonstrated by the synthesis and structural characterization of the icosahedral cluster $[Au_{13}Cl_2(PMe_2Ph)_{10}]^{3+}$.

Although credible technological and biochemical reasons may be given for studying cluster compounds of the transition metals (Muetterties 1975, 1977), our interest in these compounds derives from a more innocent desire to understand in detail the electronic and steric factors that govern the structures and stoichiometries of these complex yet aesthetically pleasing chemical species. The first examples of gold tertiary phosphine cluster compounds, namely $Au_{11}I_3(PR_3)_7$ and $[Au_9(PPh_3)_8](BF_4)_3$, were reported approximately 10 years ago by Mala-testa and Naldini (McPartlin *et al.* 1969; Bellon *et al.* 1971) and resulted from an investigation of the reactions of mononuclear gold phosphine compounds with $NaBH_4$. These compounds provided a particularly suitable vehicle for our attempt to merge synthetic and theoretical chemistry because these clusters are derived from the simple $AuPR_3$ fragment, which does not pose difficult theoretical problems because it does not have alternative stereochemical possi-bilities and the phosphine ligand does not, unlike the related carbonyl ligand, adopt alternative terminal and bridging coordination modes.

The $M(PH_3)$ fragment (M = Cu, Ni, Pd, Pt, Ag or Au) is a common constituent of cluster compounds of the platinum and Group Ia metals and has the frontier molecular orbitals represented schematically in figure 1. For these metals the five orbitals with a predominance of d orbital character are filled, and the bonding characteristics of the fragment are determined primarily by the hy(s–z) and higher-lying metal p_x and p_y orbitals. For the lighter metals, e.g. Ni or Cu, all three orbitals are used for skeletal bonding and consequently $Ni(PR_3)$ and $Cu(PR_3)$ fragments may be accurately described as being isolobal with CH or the conical

$M(CO)_3$ fragment. For the heavier elements the p_x and p_y orbitals are higher-lying and less available for bonding, and therefore the $AuPPh_3$ and $PtPPh_3$ fragments are more correctly described as being isolobal with either CH_3 or $Mn(CO)_5$ (Elian *et al.* 1976). The structural and chemical ramifications of these isolobal analogies have been discussed at some length in a recent publication (Evans & Mingos 1982).

A noteworthy limitation associated with describing $Au(PH_3)$ as isolobal with CH_3 or H can be illustrated by the recent structural determination of $Au_2(dppm-H)_2$ (Briant *et al.* 1982)

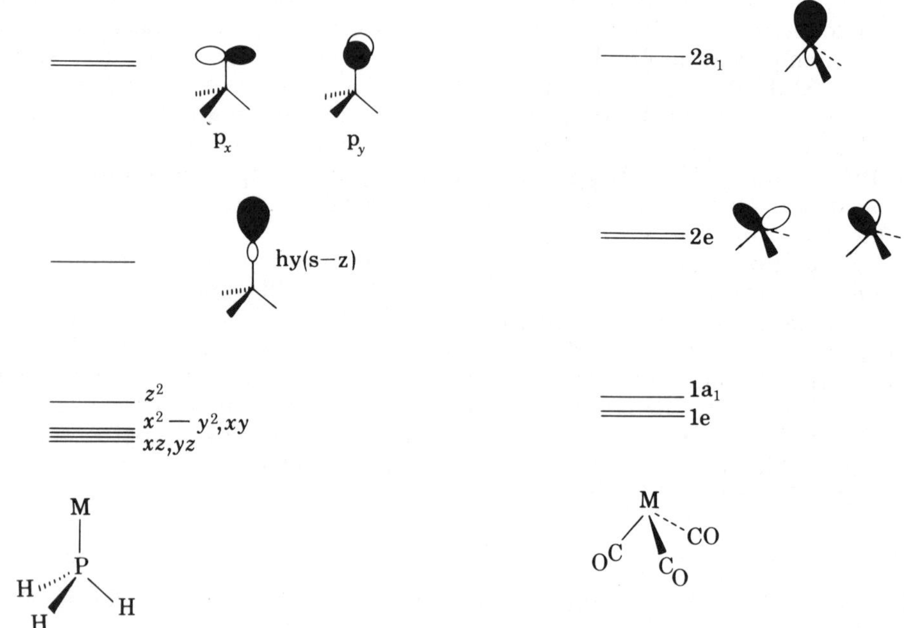

FIGURE 1. A comparison of the frontier molecule orbitals of $M(CO)_3$ and $M(PH_3)$ fragments.

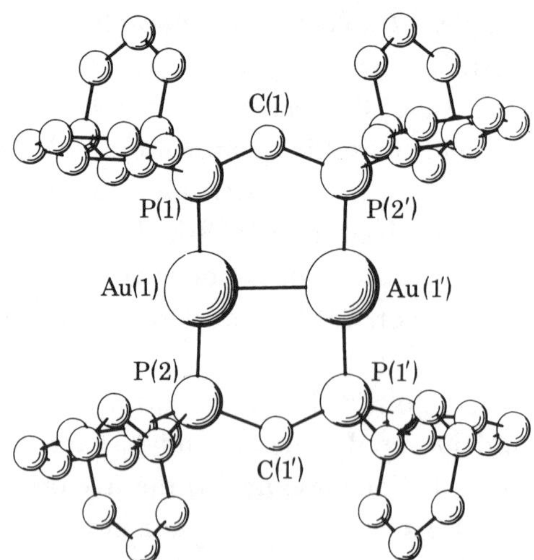

FIGURE 2. The molecular structure of $Au_2\{(Ph_2P)_2CH\}_2$. For reasons of clarity the hydrogen atoms have been omitted, and only one of the molecules that represent the total crystallographically disordered situation has been illustrated. The gold–gold distance is 2.883(3)Å.

(dppm = $Ph_2PCH_2PPh_2$). This compound resulted from the reaction of $[Au_{13}Cl_2(PMe_2Ph)_{10}]$-$(PF_6)_3$ with bis(diphenylphosphino)methane (dppm), and its molecular structure is illustrated in figure 2. In this compound the two parallel linear gold(I) (d^{10}) P–Au–P units are separated by an Au–Au distance of only 2.88 Å.† This distance, which is comparable with the gold–gold distance in the metal, can only be rationalized if the gold–gold repulsive interactions arising from the filled d orbitals on each gold atom are mitigated by the intervention of the higher-lying gold 6p orbitals, an interaction that is not available to isolobal fragments such as CH_3, $Mn(CO)_5$ or H.

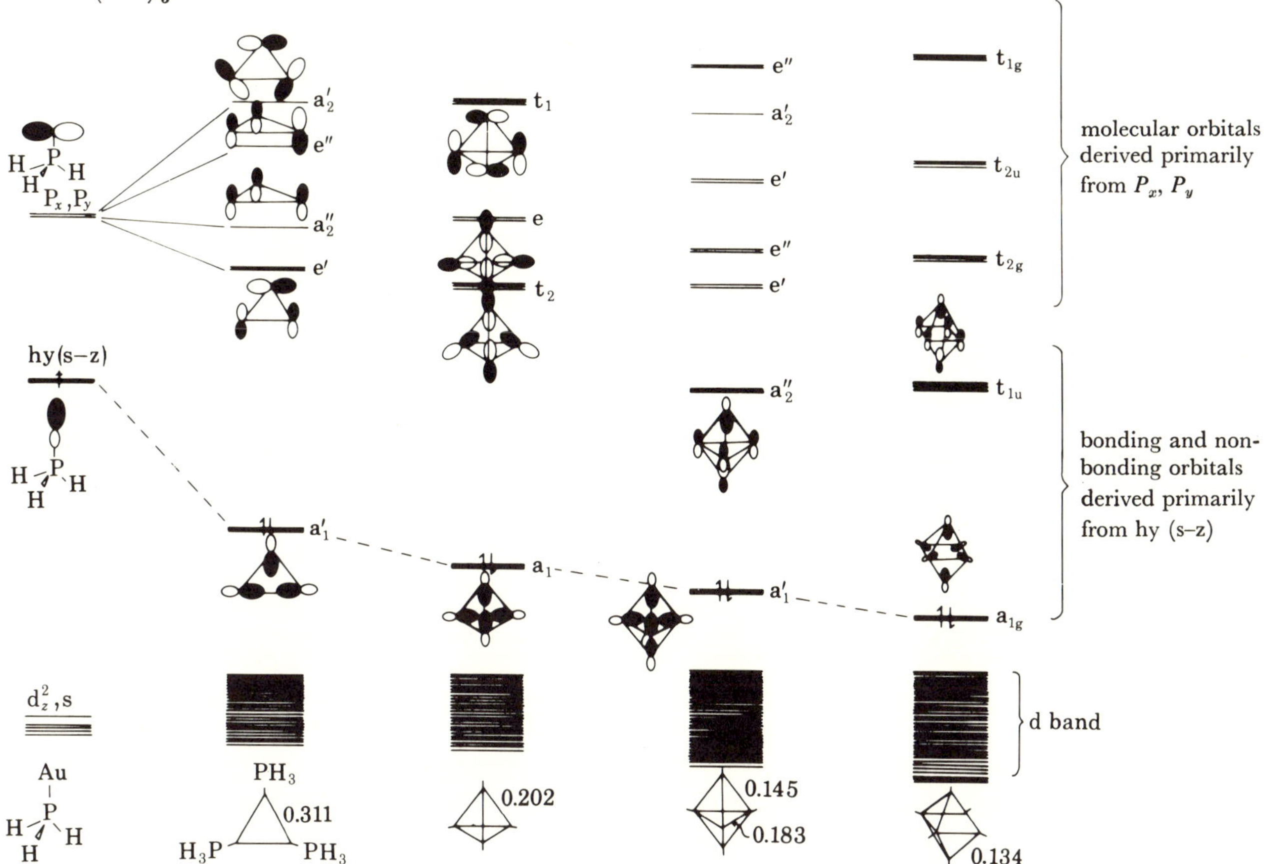

FIGURE 3. A summary of the computed bonding and non-bonding molecular orbitals of $Au_n(PH_3)_n^{x+}$ ($x = n-2$) with triangular, tetrahedral, trigonal bipyramidal and octahedral geometries. The computed Au–Au overlap populations for these ions are also given at the bottom of the figure.

The isolobal relations discussed above have been extended to homonuclear cations derived from $Au(PH_3)$ fragments (Evans & Mingos 1982) and have been used to account qualitatively for the molecular orbital diagrams illustrated in figure 3. The symmetries, nodal characteristics and relative energies of the molecular orbitals derived from $hy(s-z)$ of the $Au(PH_3)$ fragments are entirely analogous to those of H_n^{x+} aggregates, and the coefficients of the molecular orbitals may be derived from spherical harmonics in the elegant manner described by Stone (1981 a, b). From figure 3 it is apparent that this analogy is valid because the higher-lying gold 6p orbitals do not make a large direct contribution to cluster bonding.

It follows from the simple molecular orbital analysis that the cluster metal–metal bonding

† 1 Å = 10^{-10} m = 10^{-1} nm.

interactions are maximized for polyhedra with triangular faces because these generate the largest number of next-neighbour interactions between the hy(s–z) orbitals of the individual $Au(PH_3)$ fragments. The bonding may therefore be approximately described as a multicentre two-electron bond, represented in the molecular orbital terms of figure 3 as the totally symmetric (a_{1g} or a_1) filled orbital. As the nuclearity of the cluster increases, the stability of this particular molecular orbital increases (see figure 3), but the computed bond orders between individual gold atoms decreases, falling from 0.311 for the triangular cluster to 0.134 for the octahedral cluster. A suitable compromise between these opposing effects appears to be achieved for the tetrahedron and consequently this tetranuclear fragment is a common component in gold cluster and indeed mixed-metal cluster compounds containing gold. Examples of clusters containing the tetrahedral unit include $Au_4I_2(PPh_3)_4$ (Demartin *et al.* 1981), $Au_6(PPh_3)_4\{Co(CO)_4\}_2$ (Bour *et al.* 1981), $(OC)_5VAu_3(PPh_3)_3$ (Ellis 1981) and $Os_4(CO)_{13}$-$H_2Au_2(PPh_3)_2$ (Johnson *et al.* 1982).

$Au_6\{P(p\text{–}C_6H_4Me)_3\}_6^{2+}$ $Au_6(PPh_3)_6^{2+}$

FIGURE 4. Comparison of the observed bond lengths (ångströms) in $[Au_6\{(p\text{–}C_6H_4Me)_3P\}_6](BPh_4)_2$ (octahedral) and $[Au_6(PPh_3)_6](NO_3)_2$ (edge-shared tetrahedral) clusters.

In 1973 Bellon *et al.* isolated (in minute quantities) a few yellow crystals of $[Au_6\{(p\text{-tol})_3P\}_6]$-$(BPh_4)_2$ and by a single X-ray structural analysis established that it had an octahedral cluster geometry. In recent months an analogous compound $[Au_6(PPh_3)_6](NO_3)_2$ was synthesized in good yield from $[Au_8(PPh_3)_8](NO_3)_2$ and $Ag(CN)_2^-$ (Hall *et al.* 1982*b*) as a red–brown crystalline solid. Although the $^{31}P\{^1H\}$ n.m.r. spectrum of the latter showed only a single resonance at room temperature, consistent with an octahedral cluster geometry, at $-70\ ^\circ C$ the spectrum showed two resonance in the approximate ratio 2:1. A single crystal X-ray structural analysis of this compound established that its solid-state structure, namely edge-shared tetrahedral, was consistent with the low-temperature n.m.r. data. Figure 4 contrasts the observed bond lengths for these two hexanuclear gold cluster compounds. The much shorter bond lengths in the edge-shared tetrahedral cluster can be attributed to the occurrence of a pair of four-centre two-electron bonds (each associated with a single tetrahedron), rather than a single six-centre two-electron bond in the octahedron. Molecular orbital calculations (Mingos 1976) on octahedral gold clusters have shown that the second electron pair resides in a non-bonding orbital of t_{1u} symmetry. The dispersion of gold–gold bond lengths in the edge-shared tetrahedral structure can be attributed to the nodal characteristics of the highest occupied skeletal molecular orbital of b_{1u} symmetry illustrated in scheme 1. Since this orbital is strongly bonding along those metal–metal edges that are parallel and less bonding along the remaining edges an asymmetry in the computed Mulliken overlap populations (also shown in scheme 1) results.

The isolation of two clusters such as $[Au_6\{(p\text{-tol})_3P\}_6](BPh_4)_2$ and $[Au_6(PPh_3)_6](NO_3)_2$ with almost identical stoichiometries yet markedly different physical and structural properties raises many interesting questions about the relative thermodynamic stabilities of the alternative structural forms, the size of the kinetic barrier connecting them and the origin of their different physical properties. However, in the absence of a high-yield synthetic route to the octahedral cluster compound it is difficult to establish the answers to these interesting questions. For higher-nuclearity clusters we have established that apparently small changes in the phosphine ligands can result in gross changes in the polyhedral cluster geometries, and this aspect will be discussed in more detail below.

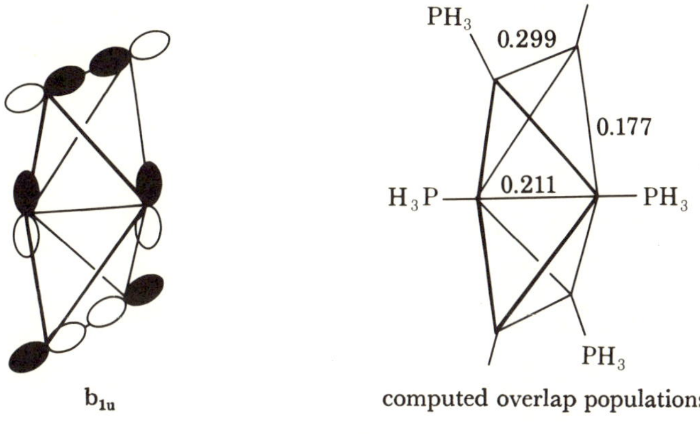

SCHEME 1

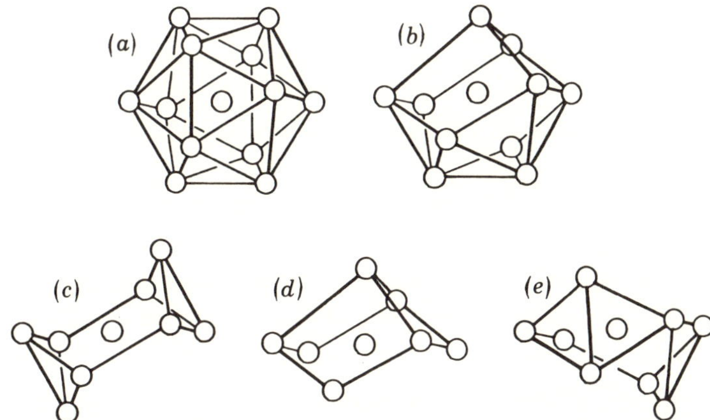

FIGURE 5. Some examples of high-nuclearity gold cluster compounds. For reasons of clarity only the cluster gold atoms have been illustrated. The compounds are: (a) $[Au_{13}Cl_2(PMe_2Ph)_{10}](PF_6)_3$; (b) $Au_{11}I_3(P(p\text{-}C_6H_4F)_3)_7$; (c) $[Au_9\{(p\text{-}C_6H_4CH_3)_3\}_8](PF_6)_3$; (d) $[Au_8(PPh_3)_8]X_2$ and (e) $Au_9(SCN)_3(PCy_3)_5$.

Higher-nuclearity cluster compounds of gold cannot be stabliized solely by the formation of multicentre bonds between the gold atoms on the periphery of the cluster, and consequently their stabilities are enhanced substantially by the presence of a central gold atom at the origin of the cluster, which can form effective radial metal–metal bonds with all the peripheral gold atoms. Some examples of such gold cluster compounds are illustrated in figure 5.

Extended Hückel molecular orbital calculations, which we have completed on a wide range of high-nuclearity gold cluster cations of the type $[Au_xL_{x-1}]^{m+}$ (Gilmour & Mingos 1982), where L is a neutral two-electron ligand, have demonstrated that the stoichiometries of these cations are decided primarily by the topology of the polyhedron defined by the peripheral gold atoms, a situation that arises from the high proportion of s character in the hy(s–z) orbitals of the $Au(PH_3)$ fragments and the spherical harmonic nature of the resultant bonding problem.

If the peripheral gold atoms define a closed spherical polyhedron, then the cluster molecular orbitals can be represented, in order of decreasing stabilities, as S, P, D functions etc. (Stone 1981). When $x = 7$–13 only the S and P functions are bonding and therefore in such complexes there are only four bonding skeletal molecular orbitals corresponding to S, P_x, P_y and P_z $(x-1)$ Au–L bonding molecular orbitals and a d-band with $5x$ molecular orbitals, resulting in a filled closed shell and stable electronic configuration when $12x+6$ valence electrons are present. Interactions between the S and P molecular orbital functions derived from the hy(s–z) orbitals of the $Au(PR_3)$ fragments and the 6s and 6p atomic wavefunctions of the central gold atom reinforce this bonding pattern to an important extent (Mingos 1976). $Au_{13}Cl_2(PMe_2Ph)_{10}^{3+}$ (162 valence electrons), $Au_{11}I_3(PPh_3)_7$ (138 valence electrons) confirm the reliability of this generalization (Briant et al. 1981; McPartlin et al. 1969).

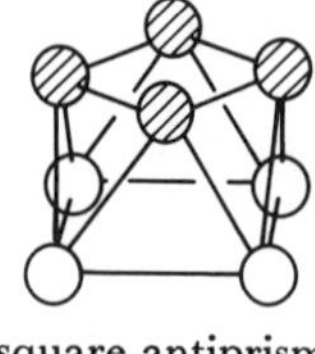
square antiprism

puckered crown

Scheme 2

If the peripheral gold atoms have a topology related to a ring or torus, then only the spherical harmonic solutions corresponding to S, P_x and P_y functions are bonding and the P_z function is either non-bonding or anti-bonding. The consequence of this topological distinction on the molecular orbitals of the peripheral gold atoms may be appreciated by a comparison of the P_z functions for a square antiprism (spherical topology) and a puckered crown of eight gold atoms (torus topology) shown in scheme 2. The wavefunction is bonding in the former case and anti-bonding in the latter. It follows that for gold cluster cations with torus topology for the peripheral atoms the total electron count is $12x+4$ valence electrons. $Au_9(PPh_3)_8^{3+}$ (112 valence electrons) and $Au_8(PR_3)_7^{2+}$ (100 valence electrons) provide examples of such open cluster species (Bellon et al. 1971; Cooper et al. 1980; Hall et al. 1982a; Manassero et al. 1979; Vollenbroek et al. 1979; van der Velden et al. 1981).

The potential energy surface for the interconversion of the alternative polyhedral forms that conform to the topological requirements outlined above is calculated to be soft, because the radial Au–Au interactions are energetically more significant than the peripheral Au–Au interactions. For the cluster compounds with torus topology the computed energy differences separating the alternative cluster geometries are sufficiently small for ligand packing and crystal packing effects to exert a dominant influence on the observed structures. The reliability of these conclusions has been demonstrated by the structural determinations which have been completed on the related cations $[Au_9\{P(p\text{-}C_6H_4Me)_3\}_8](PF_6)_3$ and $[Au_9\{P(p\text{-}C_6H_4OMe)_3\}_8]$-$(BF_4)_3$ (Bellon et al. 1971; Hall et al. 1982a, b). The cluster geometries for these two very closely related cations, which are illustrated in figure 6, show that the replacement of a methyl substituent in the *para* position of the phenyl rings by a methoxy group has led to a dramatic difference in the cluster geometries. The cluster geometry in the former may be described in terms of a centred icosahedron from which four peripheral gold atoms have been removed, but the latter compound provides a unique example of a cluster compound derived by placing a metal atom at the centre of a puckered ring (or crown) of like gold atoms. The X-ray structural analysis of $Au_9(SCN)_3(PCy_3)_5$ (Cooper *et al.* 1980) has demonstrated an alternative cluster geometry for this neutral cluster, which may also be derived from the icosahedron by atom removal and bond breaking processes. The structure of this compound is also illustrated in figure 6.

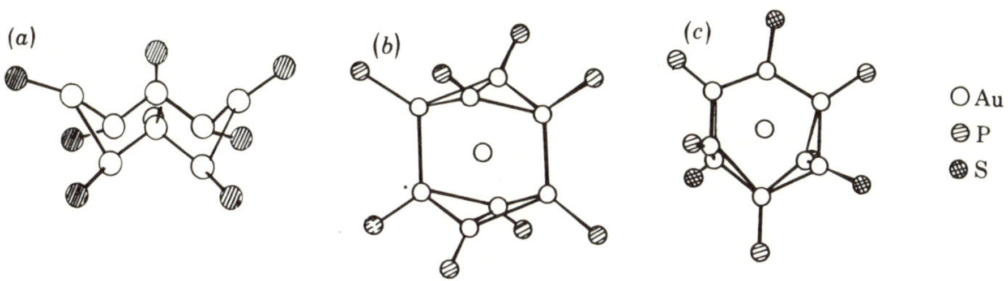

FIGURE 6. Comparison of the structures of (a) $Au_9\{P(p\text{-}C_6H_4OMe)_3\}_8^{3+}$, (b) $Au_9\{P(p\text{-}C_6H_4Me)_3\}_8^{3+}$ and (c) $Au_9(SCN)_3(PCy_3)_5$.

A comparison of the alternative structures for the Au_9 polyhedra illustrated in figure 6 also supports a previous conclusion derived from the molecular orbital calculations about the relative stabilities of the radial and tangential gold–gold interactions. Although all the compounds illustrated in figure 6 have the same number of radial gold–gold bonds, $[Au_9\{P(p\text{-}C_6H_4OMe)_3\}_8](BF_4)_3$ has four fewer tangential gold–gold bonds, which clearly suggests that the radial bonding is more significant energetically than the tangential bonding (Hall *et al.* 1982*a, b*). Mössbauer and X-ray photoelectron spectral studies have also been applied to this problem, although the conclusions have been less clearcut (Battistoni *et al.* 1977; Vollenbroek *et al.* 1978).

Solution $^{31}P\{^1H\}$ n.m.r. studies have also indicated that the gold cluster compounds with torus topologies are highly fluxional and undergo very rapid cluster rearrangement processes (Vollenbroek *et al.* 1978). However, the absence of limiting n.m.r. spectral data and the absence of a suitable nuclear spin on gold have limited such studies. Furthermore, although $[Au_9\{P(p\text{-}C_6H_4Me)_3\}_8](PF_6)_3$ and $[Au_9\{P(p\text{-}C_6H_4OMe)_3\}_8](BF_4)_3$ have very different electronic spectral characteristics in the solid state (see figure 7 for example), in solution the spectra are similar. This indicates that the cluster compounds share a common polyhedral structure in the solution phase (Hall *et al.* 1982*a*).

For gold cluster compounds with spherical topologies the computed energy differences between alternative polyhedral geometries are somewhat larger, and the most stable polyhedron is that which maximizes the number of next-nearest neighbours, e.g. the icosahedron for the $Au_{13}L_{12}^{5+}$ cluster cation. It is perhaps pertinent to indicate that the prediction of such a cluster cation (Mingos 1976) and its subsequent synthesis and structural characterization (Briant *et al.* 1981) represents by far the most important vindication of our theoretical approach. However, this prediction was made naïvely without the full realization of how small the energy differences were between the predicted icosahedron and the alternative possible polyhedral geometries. This only became apparent when n.m.r. studies on $[Au_{13}Cl_2(PMe_2Ph)_{10}]^{3+}$

6

(Briant *et al.* 1981) indicated that the structure observed in the solid state with the chloro ligands occupying *para* positions on the icosahedral cluster was not maintained in solution. The $^{31}P\{^1H\}$ n.m.r. studies have indicated the presence of the alternative *meta* and *ortho* isomers in addition to the *para* isomer in solution at room temperatures. It is important to appreciate how small the energy barriers separating these alternative isomers are compared with those reported for the corresponding isomers of the icosahedral carboranes $C_2B_{10}H_{12}$ (Wade 1971).

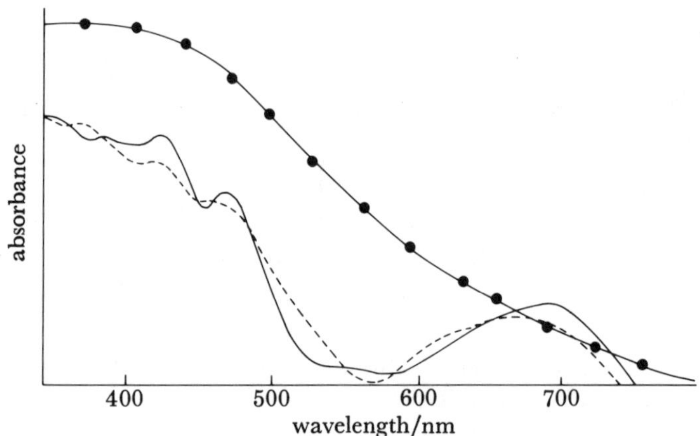

FIGURE 7. Comparison of the solid-state reflectance spectra of some $[Au_9(PR_3)_8](BF_4)_3$ cluster compounds: —, $[Au_9(PPh_3)_8](BF_4)_3$; ●—●, $[Au_9\{P(p\text{-}C_6H_4OMe)_3\}_8](BF_4)_3$; – – –, $[Au_9\{P(p\text{-}C_6H_4Me)_3\}_8](BF_4)_3$.

Although this paper has emphasized the electronic factors that govern the stoichiometries and stereochemistries of gold cluster compounds, the successful synthesis of high-nuclearity gold cluster compounds required an adequate appreciation of the steric requirements of the ligands coordinated to the peripheral gold atoms. For example, although it is possible to place eight triphenylphosphine ligands around the Au_9 torus clusters illustrated in figure 6, simple steric considerations suggested that it is not possible to arrange twelve triphenylphosphine ligands around an icosahedral gold cluster. Consequently our development of suitable synthetic routes into icosahedral gold cluster compounds utilized the tertiary phosphine ligands PMe_2Ph and $PMePh_2$, whose steric requirements are less demanding. Recently an attempt has been made to put these steric arguments onto a more quantitative basis by defining a *cluster cone angle*, which is closely related to the Tolman cone angle (Mingos 1982). This approach appears to have some utility not only for the gold cluster compounds described in this paper, but also accounting for the limiting stoichiometries in metal carbonyl cluster compounds. It will come as no surprise to chemists that cluster chemistry, like other branches of chemistry, requires the consideration simultaneously of subtle electronic and steric effects.

None of the work described above would have been possible without the dedication and experimental skills of those coworkers of mine who participated in this difficult project: Brian Theobald, James White and Kevin Hall did the synthetic studies, Clive Briant and Alan Welch the crystallographic structural determinations and David Evans and David Gilmour assisted me with the molecular orbital calculations. The S.E.R.C. is thanked for financial support, and Johnson-Matthey for a loan of gold metal.

REFERENCES

Battistoni, C., Mattogno, G., Cariati, F., Naldini, L. & Sgamellotti, A. 1977 *Inorg. chim. Acta* **24**, 207–210.

Bellon, P. L., Cariati, F., Manasserro, M., Naldini, L. & Sansoni, M. 1971 *J. chem. Soc. chem. Commun.*, pp. 1423–1424.

Bellon, P. L., Manasserro, M. & Sansoni, M. 1973 *J. chem. Soc. Dalton Trans.*, pp. 2423–2427.

Bour, J. J., van der Velden, J. W. A., Otterloo, B. F., Bosman, W. P. & Noordik, J. H. 1981 *J. chem. Soc. chem. Commun.*, p. 583.

Briant, C. E., Theobald, B. R. C., White, J. W., Bell, L. K., Mingos, D. M. P. & Welch, A. J. 1981 *J. chem. Soc. chem. Commun.*, pp. 201–202.

Briant, C. E., Hall, K. P. & Mingos, D. M. P. 1982 *J. organometall. Chem.* **229**, C5–C8.

Cooper, M. K., Dennis, G. R., Henrick, K. & McPartlin, M. 1980 *Inorg. chim. Acta* **45**, L151–L152.

Demartin, F., Manasserro, M., Naldini, L., Ruggeri, R. & Sansoni, M. 1981 *J. chem. Soc. chem. Commun.*, pp. 222–223.

Elian, M., Chen, M. M. L., Hoffmann, R. & Mingos, D. M. P. 1976 *Inorg. Chem.* **15**, 1148–1155.

Ellis, J. E. 1981 *J. Am. chem. Soc.* **103**, 6106–6110.

Evans, D. G. & Mingos, D. M. P. 1982 *J. organometall. Chem.* **232**, 171–191.

Gilmour, D. I. & Mingos, D. M. P. 1982 Unpublished.

Hall, K. P., Theobald, B. R. C., Gilmour, D. I., Mingos, D. M. P. & Welch, A. J. 1982*a* *J. chem. Soc. chem. Commun.*, pp. 528–530.

Hall, K. P., Briant, C. E. & Mingos, D. M. P. 1982*b* Unpublished.

Johnson, B. F. G., Kaner, D. A., Lewis, J., Raithby, P. R. & Taylor, M. J. 1982 *Polyhedron* **1**, 105–107.

McPartlin, M., Mason, R. & Malatesta, L. 1969 *J. chem. Soc. chem. Commun.*, pp. 334–335.

Manasserro, M., Naldini, L. & Sansoni, M. 1979 *J. chem. Soc. chem. Commun.*, pp. 385–386.

Mingos, D. M. P. 1976 *J. chem. Soc. Dalton Trans.*, pp. 1163–1169.

Mingos, D. M. P. 1982 *Inorg. Chem.* **21**, 464–466.

Muetterties, E. L. 1975 *Bull. Soc. chim. Belg.* **84**, 959.

Muetterties, E. L. 1977 *Science, Wash.* **196**, 839–848.

Stone, A. J. 1981*a* *Inorg. Chem.* **20**, 563.

Stone, A. J. 1981*b* *Molec. Phys.* **41**, 1339–1354.

Vollenbroek, F. A., Bouten, P. C. P., Trooster, J. M., van der Berg, J. P. & Bour, J. J. 1978 *Inorg. Chem.* **17**, 1345–1347.

Vollenbroek, F. A., Bosman, W. P., Noordik, J. H., Bour, J. J. & Beurskens, P. T. 1979 *J. chem. Soc. chem. Commun.*, pp. 387–388.

van der Velden, J. W. A., Bour, J. J., Bosman, W. P. & Noordik, J. H. 1981 *J. chem. Soc. chem. Commun.*, pp. 1218–1219.

Wade, K. 1971 *Electron deficient compounds*, pp. 155–160. London: Nelson & Sons.

Phil. Trans. R. Soc. Lond. A **308**, 85–86 (1982) [85]
Printed in Great Britain

General discussion

M. B. HALL (*Department of Chemistry, Texas A & M University, U.S.A.*). The chemistry of $H_2Os_3(CO)_{10}$ resembles that of an alkene because both have an empty, low-lying π^* orbital, which accepts electrons from an attacking nucleophile. However, the similarity of the unoccupied orbital does not imply a similarity of the occupied orbitals. Molecular orbital calculations and photoelectron spectra (Sherwood & Hall, *Inorg. Chem.* (in the press); Green, Mingos & Seddon, *Inorg. Chem.* **11**, 1619 (1981)) suggest that the bonding within the H_2Os_2 unit is best described as two three-centre, two-electron Os–H–Os bridging bonds similar to those in B_2H_6. Although the analogy with diborane is closer than that with ethylene, it is not completely accurate. Our calculations show that the H bridges stabilize some of the orbitals involved in Os to CO π backbonding and that this stabilization leads to some net, direct Os–Os bonding in addition to the Os–H–Os bridging bonds.

J. C. GREEN (*Inorganic Chemistry Laboratory, University of Oxford, U.K.*). The general form of the photoelectron spectrum of a metal carbonyl compound containing a metal–metal bond consists of the lowest ionization energy bands (7.4–9 eV) resulting from ionization of metal–metal bonding electrons followed by bands associated with metal d electrons involved in backbonding to the carbonyl ligands (9–11 eV) (Green *et al.* 1981). This is illustrated in figure 1 by the p.e. spectrum of $Os_3(CO)_{12}$. Thus the metal–metal bonding electrons are the easiest to remove from the cluster.

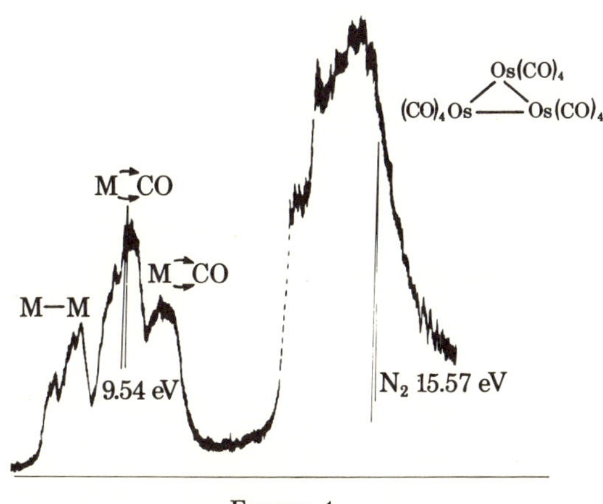

FIGURE 1

When, in an isoelectronic compound, an M–M interaction is replaced by an M–H–M bridge, the ionization band previously due to the M–M bond is absent and is replaced at considerably higher ionization energy (11.5–13 eV) by another band, which may be associated with ionization of the two electrons binding the M–H–M bridge. Both the high binding energy of these bands and their intensity characteristics indicate a considerable degree of localization on the hydrogen atom. This type of spectrum is illustrated by the p.e. spectrum of $Re_3H_3(CO)_{12}$ (figure 2).

In compounds where both M–M and M–H–M bonds are present, both types of band are found in their characteristic regions, indicating that bonding in this mixed type of cluster is fairly localized. If this type of cluster were to be subjected to electrophilic attack, such as oxidation by iodine, as Dr Mays described in his paper, that attack would be predicted to centre round the most easily removed electrons, namely the M–M bond rather than the more tightly bound M–H–M electron pair.

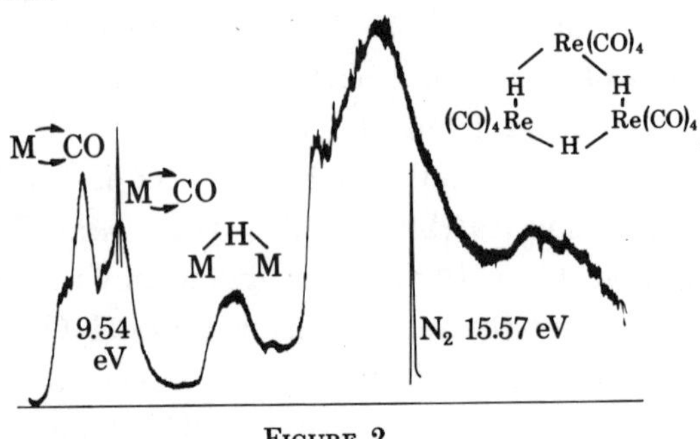

FIGURE 2

We may conclude that the role of heteroatoms is twofold. They not only provide electrons necessary for binding the cluster but they also have the capacity to stabilize markedly the cluster bonding orbitals.

Reference

Green, J. C., Mingos, D. M. P. & Seddon, E. A. 1981 *Inorg. Chem.* **20**, 2595.

Phil. Trans. R. Soc. Lond. A **308**, 87–93 (1982)　　　[87]
Printed in Great Britain

Novel routes to heteronuclear metal clusters containing rhodium

By F. G. A. STONE, F.R.S.

Department of Inorganic Chemistry, University of Bristol, Bristol BS8 1 TS, U.K.

Logical methods for preparing small metal clusters with heteronuclear metal–metal bonds can be devised by appreciating the isolobal relationships that exist between certain metal ligand combinations and organic fragments. Thus the group $W(CO)_2(\eta\text{-}C_5H_5)$ has frontier orbitals of similar pattern to those of CH, while those of $Rh(CO)(\eta\text{-}C_5Me_5)$ and CH_2 are also similar. This paper shows how such ideas have been employed to generate compounds with bonds between rhodium and other metals.

INTRODUCTION

It has long been known that certain metal ligand fragments and organic species show to some degree similar reactivity patterns (e.g. $Pt(PR_3)_2$ and CH_2). Only recently, however, has the scope of these relations been fully appreciated, and an explanation given by Hoffmann (1982) in terms of the similarity of the frontier orbitals of the inorganic and organic groups. Two fragments are isolobal if the number, symmetry properties, approximate energy and shape of the frontier orbitals, and the number of electrons in them, are similar. By reference to Hoffmann's lecture (1982), it will be seen that the CH_2 or carbene moiety is isolobal with several metal ligand systems, e.g. PtL_2, $Rh(CO)(\eta\text{-}C_5Me_5)$, $Fe(CO)_4$, $Mn(CO)_2(\eta\text{-}C_5H_5)$, $Cr(CO)_5$, $Cr(CO)_2(\eta\text{-}C_6H_6)$, formally derived by removing electron-pair donor ligands L from an octahedral complex ML_6. Similarly, a carbyne fragment CR is isolobal with the species $W(CO)_2(\eta\text{-}C_5H_5)$ (W, d^5, ML_5), $RhR(\eta\text{-}C_9H_7)$† (Rh, d^7, ML_4), and $Co(CO)_3$ (Co, d^9, ML_3). These relations allow designed syntheses of compounds with heteronuclear metal–metal bonds, rather than the *ad hoc* preparative procedures previously employed. Because of time limitations, I shall restrict discussion to compounds involving rhodium.

THE ALKYLIDYNE COMPOUND $[W{\equiv}CC_6H_4Me\text{-}4(CO)_2(\eta\text{-}C_5H_5)]$

The compound $[RC{\equiv}W(CO)_2(\eta\text{-}C_5H_5)]$ (R = C_6H_4Me-4 in the formulae) reacts with low-valent metal complexes in a similar manner to an alkyne, a property first exploited by us in reactions involving platinum.‡ Compounds **1** and **2** illustrate our early results, which subsequently led Hoffmann (1982) to point out the frontier orbital relation between $W(CO)_2(\eta\text{-}C_5H_5)$ and CR, following our intuitive ideas for synthetic work.

In the context of rhodium chemistry, the indenyl compound $[Rh(CO)_2(\eta\text{-}C_9H_7)]$ reacts with the alkylidynetungsten complex to afford **3**. Treatment of the latter with $[Fe_2(CO)_9]$ gives the trimetal cluster **4**. Isolobal relationships allow **3** to be regarded as a dimetallacyclopropene, and **4** as a trimetallatetrahedrane. Compound **4** is perhaps the first example of a cluster species involving three metals each of which belongs to a different transition element series.

† $\eta\text{-}C_9H_7$ = the indenyl ligand.
‡ References to our experimental results are given in the appendix.

Treatment of **3** with $[\text{Rh(CO)}_2(\eta\text{-C}_9\text{H}_7)]$ in toluene at *ca.* 60 °C affords the dirhodium-tungsten cluster **5**. Compound **6** is another species with a $\mu_3\text{-CRh}_2\text{W}$ core. It is prepared in a two-step synthesis, the first of which involves reacting $[\text{W}{\equiv}\text{CR(CO)}_2(\eta\text{-C}_5\text{H}_5)]$ with $[\text{Rh(acac)(CO)}_2]$ (acac = acetylacetonate) at room temperature to produce **7**. In the second

$$(\eta\text{-C}_5\text{H}_5)\,(\text{OC})_2\text{W}\!=\!\!=\!\!=\!\text{Pt(PR}_3')_2 \qquad \mathbf{1}$$

$$(\eta\text{-C}_5\text{H}_5)\,(\text{OC})\text{W} \cdots \text{Pt} \cdots \text{W(CO)}\,(\eta\text{-C}_5\text{H}_5) \qquad \mathbf{2}$$

$$(\eta\text{-C}_5\text{H}_5)\,(\text{OC})_2\text{W}\!=\!\!=\!\!=\!\text{Rh(CO)}\,(\eta\text{-C}_9\text{H}_7) \qquad \mathbf{3}$$

$$(\eta\text{-C}_5\text{H}_5)\,(\text{OC})_2\text{W} \cdots \text{Rh}(\eta\text{-C}_9\text{H}_7);\ \ \text{Fe(CO)}_3 \qquad \mathbf{4}$$

$$(\eta\text{-C}_5\text{H}_5)\,(\text{OC})_2\text{W} \cdots \text{Rh}(\eta\text{-C}_9\text{H}_7);\ \ \text{Rh}(\eta\text{-C}_9\text{H}_7) \qquad \mathbf{5}$$

$$(\eta\text{-C}_5\text{H}_5)\,(\text{OC})_2\text{W} \cdots \text{Rh (acac)};\ \ \text{Rh (acac)} \qquad \mathbf{6}$$

$$(\eta\text{-C}_5\text{H}_5)\,(\text{OC})_2\text{W}\!=\!\!=\!\!=\!\text{Rh(acac)(CO)} \qquad \mathbf{7}$$

step, **7** is treated with $[\text{Rh(acac)(C}_2\text{H}_4)_2]$. The latter readily releases ethylene, so that a Rh(acac) fragment adds to the dimetallacyclopropene ring system, yielding the trimetal compound **6**. The principles underlying the preparation of **4–6** have been applied to several other systems so that these processes provide a general route to compounds having $\mu_3\text{-CM}'\text{M}''\text{W}$ or $\mu_3\text{-CM}_2\text{W}$ core structures.

THE COMPOUND $[\text{Rh}_2(\mu\text{-CO})_2(\eta\text{-C}_5\text{Me}_5)_2]$

Jones *et al.* (1978) found that reduction of $[\text{Rh(CO)}_2(\eta\text{-C}_5\text{H}_5)]$ with sodium amalgam in tetrahydrofuran afforded, *inter alia*, the trirhodium anion $[\text{Rh}_3(\mu_3\text{-CO})_2(\text{CO})_2(\eta\text{-C}_5\text{H}_5)_2]^-$. This observation stimulated Pinhas *et al.* (1980) to examine the bonding in the anion in relation to its structure. Several interesting results came from this study. The trirhodium species may be regarded as an adduct of the unknown molecule $[\text{Rh}_2(\mu\text{-CO})_2(\eta\text{-C}_5\text{H}_5)_2]$ (isolobal with ethylene) and $[\text{Rh(CO)}_2]^-$ (isolobal with CH_2). Moreover, $[\text{Rh}_3(\mu_3\text{-CO})_2(\text{CO})_2(\eta\text{-C}_5\text{H}_5)_2]^-$

is clearly related to Herrmann's (1982) carbene bridged complexes $[Rh_2(\mu\text{-}CR_2)(CO)_2(\eta\text{-}C_5H_5)_2]$. Furthermore the latter and the trirhodium anion have similar ring-bonding to that of cyclopropane.

Whereas $[Rh_2(\mu\text{-}CO)_2(\eta\text{-}C_5H_5)_2]$ has not yet been isolated, its analogue $[Rh_2(\mu\text{-}CO)_2(\eta\text{-}C_5Me_5)_2]$ has been prepared by Nutton & Maitlis (1979). Moreover, the $\eta\text{-}C_5Me_5$ and $\eta\text{-}C_5H_5$ derivatives are similar in formally containing $Rh{=}Rh$ double bonds. Or, more precisely (Pinhas et $al.$ 1980), both molecules have a $2b_2$ orbital like the π^* orbital of ethylene, and available for acceptance of electrons from a nucleophilic fragment.

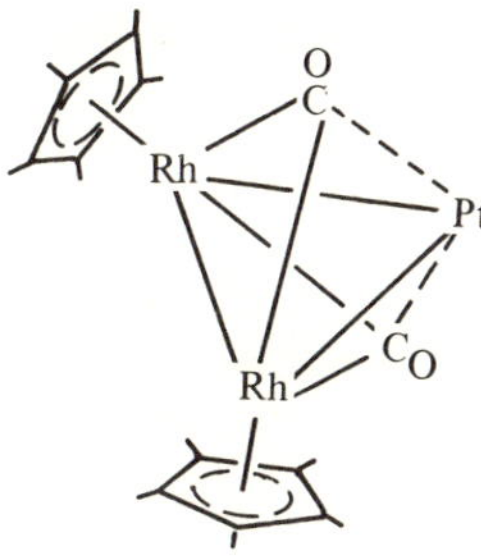

	L	L'
8a	cyclo-C_8H_{12}	
8b	CO	PPh$_3$
8c	CO	CO

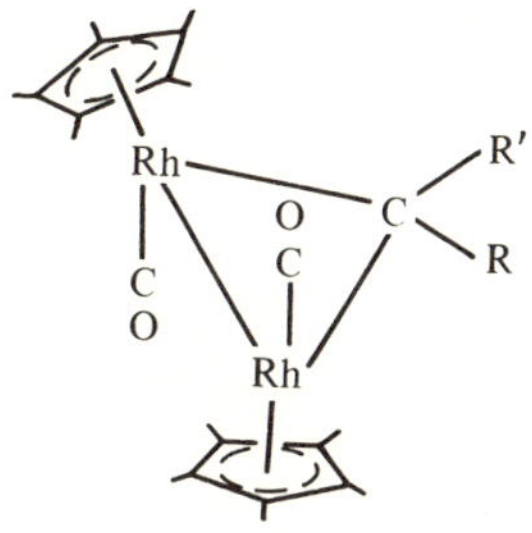

	R	R'
9a	H	H
9b	CF$_3$	CF$_3$
9c	H	CH:CH$_2$

The isolobal relationships existing between $[Rh_2(\mu\text{-}CO)_2(\eta\text{-}C_5Me_5)_2]$ and C_2H_4, or between $Rh(CO)(\eta\text{-}C_5Me_5)$ and CH_2, have guided us in the preparation of several complexes with bonds between rhodium and other transition metals. Thus $[Rh_2(\mu\text{-}CO)_2(\eta\text{-}C_5Me_5)_2]$ would be expected to combine with those metal fragments (e.g. $Pt(PR_3)_2$, $Fe(CO)_4$) that bond ethylene, and trimetal complexes should exist containing a $Rh(CO)(\eta\text{-}C_5Me_5)$ fragment bridging metal–metal bonds that without this added moiety would formally have double-bond character.

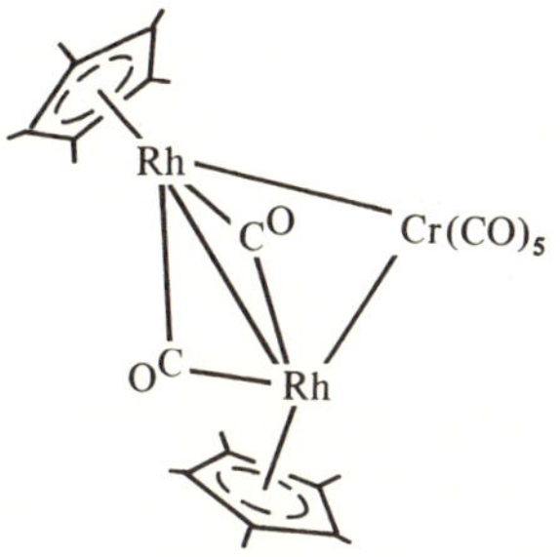

10

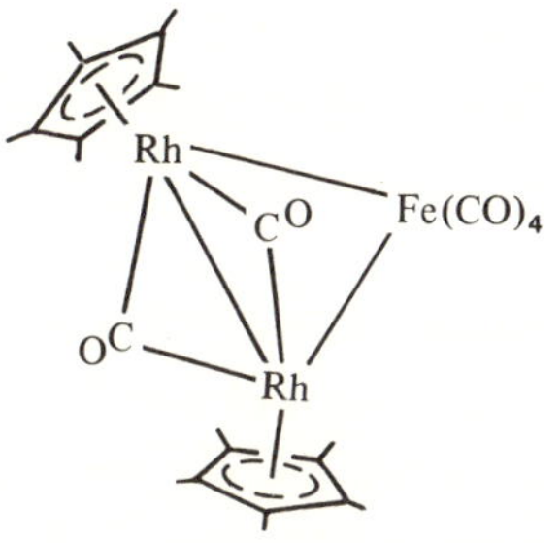

11

In the context of these ideas, the dirhodiumplatinum compounds 8 have been prepared by treating $[Rh_2(\mu\text{-}CO)_2(\eta\text{-}C_5Me_5)_2]$ with low-valent platinum complexes. In related work it was shown that the dirhodium compound reacts with diazoalkanes to afford the bridged alkylidene complexes 9. Compounds 8 and 9 may be regarded as trimetalla- and dimetalla-cyclopropanes, respectively. Related to 8 are the trimetal compounds 10 and 11, both obtained from $[Rh_2(\mu\text{-}CO)_2(\eta\text{-}C_5Me_5)_2]$ by addition of the carbene-like fragments $Cr(CO)_5$ and $Fe(CO)_4$, respectively. In compounds 8–11, the CO ligands are either terminally bound (9), bridge two metal centres (10), or asymmetrically bridge three metal atoms (8). However, it is important to recognize that transformations between bridging and non-bridging carbonyl groups, a common feature of metal carbonyl chemistry, do not cause a major perturbation of the nature of the frontier orbitals in the metal fragments (Hoffmann 1982). Indeed, in solution **11** exists as a mixture of tautomers, involving an equilibrium between molecules with the structure shown and others with two semi-bridging and one triply-bridging CO ligands.

Perhaps the most spectacular example of the similarity in the coordination behaviour of $[Rh_2(\mu\text{-}CO)_2(\eta\text{-}C_5Me_5)_2]$ and ethylene is provided by the pentanuclear metal complex $[Pt\{Rh_2(\mu\text{-}CO)_2(\eta\text{-}C_5Me_5)_2\}_2]$ (**12**) obtained by reacting the dirhodium compound with $[Pt(C_2H_4)_3]$. Two crystalline forms (monoclinic and orthorhombic) of **12** have been studied

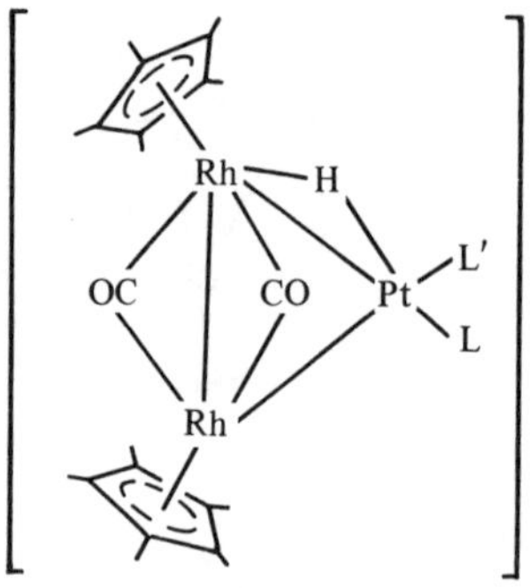

L	L'
cyclo-C_8H_{12}	
CO	CO
CO	PPh_3

13

by X-ray diffraction. In both forms the four rhodium atoms are disposed around the platinum in an essentially tetrahedral environment, with the dihedral angles between the PtRh$_2$ planes being 90° (orthorhombic) and 100° (monoclinic). The four CO ligands, in addition to bridging the Rh–Rh vectors, weakly interact with the platinum atom; a feature that persists in solution, as deduced from ^{195}Pt–^{13}C coupling observed in the ^{13}C n.m.r. spectrum.

No X-ray diffraction studies have been made on molecules $[M(C_2H_4)_2]$, since they have only a fleeting existence. However, a family of stable compounds $[Pt(alkyne)_2]$ is known (Boag *et al.* 1980) and have structures in which the four ligated carbon atoms adopt a pseudo-D_{2d} arrangement around the platinum atom, as do the four rhodium atoms in **12**. For the hypothetical molecule $[Ni(C_2H_4)_2]$, calculations (Rösch & Hoffmann 1974) indicate only a small energy difference between D_{2d} and D_{2h} structures, because the ethylene π^* and metal d orbitals are far apart in energy and non-degenerate perturbation theory predicts similar stabilization. However, where the energy separation between the orbitals is small, as might be expected for the $2b_2$ orbitals of the two $[Rh_2(\mu\text{-}CO)_2(\eta\text{-}C_5Me_5)_2]$ moieties and the 5d orbitals of platinum, there would be a preference for a D_{2d} structure, as is observed. However, too much reliance should not be placed on these considerations because D_{2h} symmetry for the core metal atoms of **12** would result in steric crowding of the $\eta\text{-}C_5Me_5$ groups.

Protonation of the compounds **8** with HBF$_4$ in diethyl ether affords the salts **13**. These are interesting in several respects. In the context of isolobal relationships they are similar to edge-protonated cyclopropanes, the $[Pt(H)L_2]^+$ fragment being isolobal with CH_3^+. The salts **13** have structures in which the CO ligands bridging the Rh–Rh vectors do not triply bridge to the platinum to the same degree as do the corresponding CO groups in the neutral species **8**. In solution, the salts undergo dynamic behaviour involving rotation of the $Rh_2(\mu\text{-}CO)_2(\eta\text{-}C_5Me_5)_2$ fragment about an axis through the platinum and the midpoint of the Rh=Rh unit, while the $Pt(H)L_2$ moiety retains its integrity at all times.

THE RHODIUM FRAGMENT $Rh(CO)(\eta\text{-}C_5Me_5)$

The species $Rh(CO)(\eta\text{-}C_5Me_5)$ is isolobal with CH_2, because it is effectively a fragment RhL_4 (Rh^I, d^8) like $Fe(CO)_4$ (Fe^0, d^8) (Hoffmann 1982). Hence the existence of the compound **14** is not surprising; a trimetal cluster we have obtained from the reaction between $[Rh(CO)_2(\eta\text{-}C_5Me_5)]$ and $[Fe_2(CO)_9]$ in tetrahydrofuran (thf). It is interesting to compare **14** with the isolobally related methylene adduct of the $Fe_2(CO)_8$ molecule, namely $[Fe_2(\mu\text{-}CH_2)(CO)_8]$ (Sumner 1980).

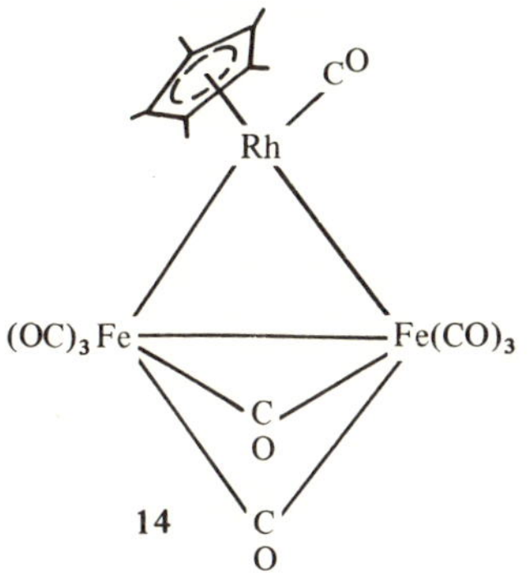

14 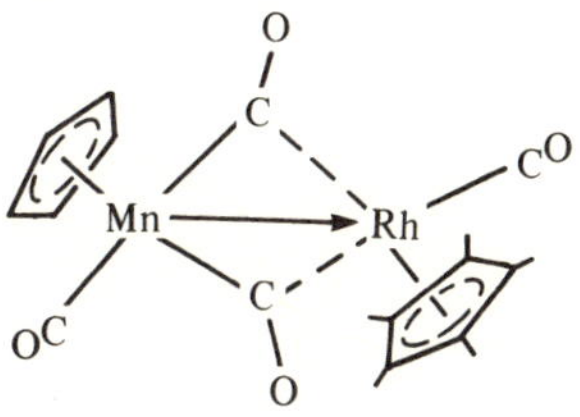   **15**

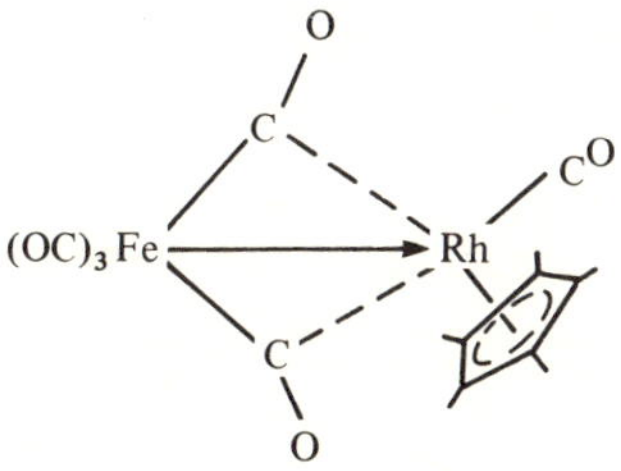

16 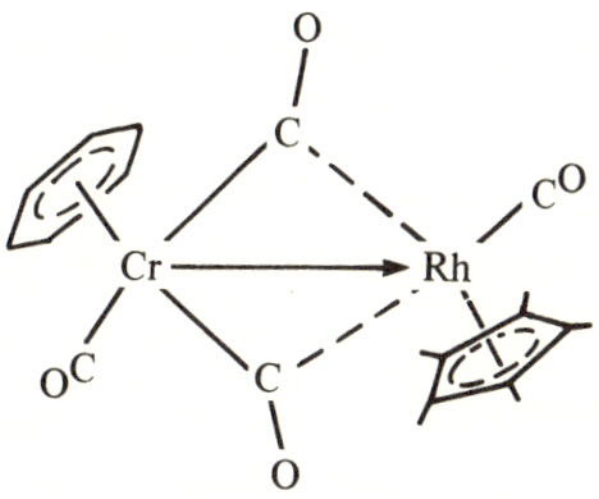   **17**

The compound $[Rh(CO)_2(\eta\text{-}C_5Me_5)]$ undergoes an interesting group of reactions with the species $[Mn(CO)_2(thf)(\eta\text{-}C_5H_5)]$, $[Fe(CO)_4(thf)]$ and $[Cr(CO)_2(thf)(\eta\text{-}C_6H_6)]$. By release of the weakly coordinated thf molecules, these three reagents afford the carbene-like metal–ligand moieties $Mn(CO)_2(\eta\text{-}C_5H_5)$, $Fe(CO)_4$ and $Cr(CO)_5$. The products of the three reactons are the bimetallic compounds **15–17**, the formation of which has resulted in the transfer of a CO group from rhodium to Mn, Fe and Cr, respectively. Although not metal clusters, in the normal definition, the complexes **15–17** illustrate the isolobal concept, which so usefully links inorganic and organic chemistry. All three compounds contain the $Rh(CO)(\eta\text{-}C_5Me_5)$ fragment to which an electron pair has been donated from the h.o.m.o. of the molecules $[Mn(CO)_3(\eta\text{-}C_5H_5)]$, $[Fe(CO)_5]$ and $[Cr(CO)_3(\eta\text{-}C_6H_6)]$, an orbital that is probably metal-centred (Lichtenberger & Fenske 1976). Consequently, **15–17** are akin to ylides $R_3P \rightarrow CH_2$: in the three metal complexes excess charge on the rhodium is dissipated via the semi-bridging carbonyls, whereas with ylides the canonical form $R_3P{=}CH_2$ is important.

FUTURE PROSPECTS

The above brief account of a small portion of our recent work will, I hope, indicate the usefulness of approaching the synthesis of compounds with metal–metal bonds by giving due regard to the frontier orbital similarities of the metal ligand fragments involved. The potential of this approach is illustrated by two very recent researches, which serve to close this discussion.

Again by reference to Hoffmann (1982) it will be seen that a T-shaped ML_3 (M, d^8) fragment is isolobal with CH_3^+ or CH_2. Such fragments are available for Rh^I or Ir^I by release of $NH_2C_6H_4Me$-4 from the complexes $[MCl(CO)_2(NH_2C_6H_4Me$-4$)]$ (M = Rh or Ir). Accordingly, reactions have been studied between the latter and $[Os_3(\mu$-H$)_2(CO)_{10}]$, a molecule

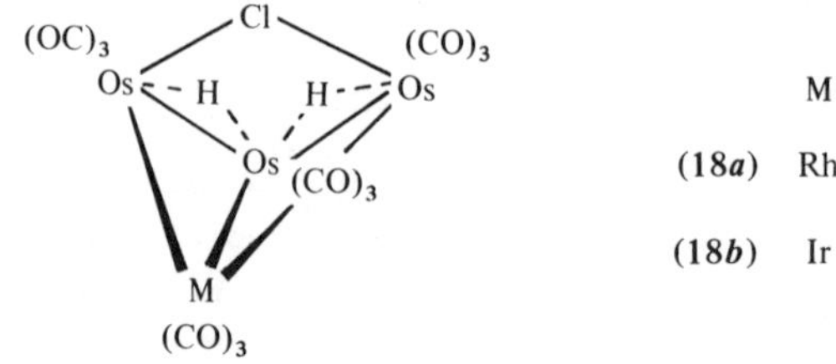

known to add carbene groups readily. The products obtained, **18**, have a 'butterfly' arrangement of the metal atom core, with the wingtips bridged by Cl and hydrido ligands in the positions indicated. These compounds do not contain T-shaped $MCl(CO)_2$ 'carbene' fragments, but $M(CO)_3$ groups isolobal with CH. This prompts a cautionary comment about isolobal mapping. There is no guarantee that a synthesis based on a correlation between the isolobal nature of two groups (e.g. $RhCl(CO)_2$ and CH_2) will result in products with similar structures. Kinetic and thermodynamic factors will influence the structure of the final products.

We have stressed earlier the usefulness of $[Rh_2(\mu$-CO$)_2(\eta$-$C_5Me_5)_2]$ as a building block in metal cluster chemistry. The cobalt analogue is known, and a good method of synthesis has recently been developed (Beevor 1981). This molecule has been shown (Cirjak *et al.* 1980) to add a number of photolytically generated metal fragments, thereby affording a variety of heteronuclear trimetal complexes. Clearly this important work is related to that discussed in this paper.

Of particular interest would be mixed-metal species containing formally Co=Rh bonds, because addition of other metal ligand fragments to these molecules should afford tri- or tetra-nuclear clusters with three different transition elements. With this in mind, what is conceptually evident is the possible dimerization of $Co(CO)(\eta$-$C_5Me_5)$ and $Rh(CO)(\eta$-$C_5Me_5)$ fragments to give $[CoRh(\mu$-CO$)_2(\eta$-$C_5Me_5)_2]$ (**19**). This we have accomplished by reacting $[Co(C_2H_4)_2(\eta$-$C_5Me_5)]$ with $[Rh(CO)_2(\eta$-$C_5Me_5)]$ because the former can scavenge a CO group from the latter. Compound **19** has recently been shown to combine with several 'carbene-like' metal–ligand fragments, e.g. $Mo(CO)_5$, $Fe(CO)_4$, and $Pt(cyclo$-$C_8H_{12})$, to afford trimetal clusters.

REFERENCES

Beevor, R. G., Frith, S. A. & Spencer, J. L. 1981 *J. organometall. Chem.* **221**, C25.
Boag, N. M., Howard, J. A. K., Green, M., Grove, D. M., Spencer, J. L. & Stone, F. G. A. 1980 *J. chem. Soc. Dalton Trans.*, p. 2170.
Cirjak, L. M., Huang, J.-S., Zhu, Z.-H. & Dahl, L. F. 1980 *J. Am. chem. Soc.* **102**, 6623.
Herrmann, W. A. 1982 *Adv. organometall. Chem.* **20**, 159.
Hoffmann, R. 1982 *Les Prix Nobel 1981.* Stockholm: Almqvist & Wiksell.

Jones, W. D., White, M. A. & Bergman, R. D. 1978 *J. Am. chem. Soc.* **100**, 6770.
Lichtenberger, D. L. & Fenske, R. F. 1976 *J. Am. chem. Soc.* **98**, 50.
Nutton, A. & Maitlis, P. M. 1979 *J. organometall. Chem.* **166**, C21.
Pinhas, A. R., Albright, T. A., Hofmann, P. & Hoffmann, R. 1980 *Helv. chim. Acta* **63**, 29.
Rösch, N. & Hoffmann, R. 1974 *Inorg. Chem.* **13**, 2656.
Sumner, C. E., Riley, P. E., Davis, R. E. & Pettit, R. 1980 *J. Am. chem. Soc.* **102**, 1752.

Appendix

The articles listed below describe our own experimental results reviewed in this symposium.

Compounds **1** *and* **2**:
Ashworth, T. V., Howard, J. A. K. & Stone, F. G. A. 1979 *J. chem. Soc. chem. Commun.*, p. 42.
Ashworth, T. V., Howard, J. A. K. & Stone, F. G. A. 1980 *J. chem. Soc. Dalton Trans.*, p. 1609.
Ashworth, T. V., Chetcuti, M. J., Howard, J. A. K., Stone, F. G. A., Wisbey, S. J. & Woodward, P. 1981 *J. chem. Soc. Dalton Trans.*, p. 763.

Compounds **3–5**:
Green, M., Jeffery, J. C., Porter, S. J., Razay, H. & Stone, F. G. A. 1982 *J. chem. Soc. Dalton Trans.* (In the press.)

Compounds **6** *and* **7**:
Chetcuti, M. J., Chetcuti, P. A. M., Jeffery, J. C., Mills, R. M., Mitrprachachon, P., Pickering, S. J., Stone, F. G. A. & Woodward, P. 1982 *J. chem. Soc. Dalton Trans.*, p. 699.

Compounds **8** *and* **9**:
Green, M., Mills, R. M., Pain, G. N., Stone, F. G. A. & Woodward, P. 1982 *J. chem. Soc. Dalton Trans.*, p. 1309.

Compound **10**:
Barr, R. D., Green, M. & Stone, F. G. A. *J. chem. Soc. Dalton Trans.* (Submitted.)

Compounds **11, 14, 15** *and* **16**:
Aldridge, M. L., Green, M., Howard, J. A. K., Pain, G. N., Porter, S. J., Stone, F. G. A. & Woodward, P. 1982 *J. chem. Soc. Dalton Trans.*, p. 1333.

Compound **12**:
Green, M., Howard, J. A. K., Pain, G. N. & Stone, F. G. A. 1982 *J. chem. Soc. Dalton Trans.*, p. 1327.

Compound **13**:
Green, M., Mills, R. M., Pain, G. N., Stone, F. G. A. & Woodward, P. 1982 *J. chem. Soc. Dalton Trans.*, p. 1321.

Compound **17**:
Barr, R. D. Bristol University, unpublished results.

Compound **18**:
Farrugia, L. J., Orpen, A. G. & Stone, F. G. A. 1982 *Polyhedron.* (In the press.)

Compound **19**:
Hankey, D. R. Bristol University, unpublished results.

Phil. Trans. R. Soc. Lond. A **308**, 95–102 (1982)   [95]
Printed in Great Britain

Nuclear magnetic resonance studies in cluster chemistry

By B. T. Heaton

The Chemical Laboratory, University of Kent at Canterbury, Canterbury, Kent CT2 7NH, U.K.

Multinuclear n.m.r. studies on transition metal carbonyl clusters are providing important information about (*a*) mechanisms of ligand inter- and intra-exchange, (*b*) metal polyhedral rearrangements, (*c*) the controlled capping of Rh_3-triangular and Rh_4-square faces, which often provides a convenient route for the synthesis of higher nuclearity carbonyl clusters, and (*d*) bonding within the cluster. Recent ^{13}C and ^{31}P n.m.r. measurements under high pressures of gas promise *in situ* mechanistic information and structural identification of intermediates present under industrially forcing conditions that are usually necessary for catalysis.

Introduction

The development of X-ray structural analysis over the last few years, coupled with the preparative expertise of cluster chemists in Milan, Cambridge and Union Carbide, has allowed the characterization of increasingly bigger homometallic transition metal carbonyl clusters, which now start to resemble small metallic crystallites. Similar advances are being made with alloy-like, heterometallic clusters. The advent of Fourier transform multi-n.m.r. spectrometers enabled the improved structural characterization in the solid state to be paralleled in solution; additionally, n.m.r. measurements on rhodium carbonyl clusters have allowed important information to be obtained about (*a*) ligand and metal inter- and intra-exchange processes, (*b*) reactivity and capping of particular metal polyhedral faces, which resemble surface reconstruction on metals, and (*c*) bonding within the cluster. Recently it has been possible to carry out n.m.r. measurements under high pressures of gas, which allows mechanistic information to be obtained *in situ* under the industrially forcing conditions that are usually necessary for catalysis. Coincidentally, rhodium carbonyl clusters, which appear to be most active catalytically, are best suited for n.m.r. studies because rhodium exists totally as ^{103}Rh with $I = \frac{1}{2}$, and allows this allows spin–spin coupling information, which assists with structural assignments and mechanistic information relating to inter- or intra-ligand exchange processes to be obtained. Additionally, it is now possible to obtain direct ^{103}Rh n.m.r. spectra on high-field spectrometers and these, together with $^{13}C-\{^{103}Rh\}$ and $^{1}H-\{^{103}Rh\}$ measurements, usually allow unambiguous spectroscopic assignments.

Mechanisms of ligand–metal exchange from n.m.r. studies

(*a*) Carbonyl inter-exchange

It has recently been shown (Heaton *et al.* 1981*b*) that preferential CO inter-exchange in $[Rh_6(CO)_{15}C]^{2-}$ occurs at sites that have the longest Rh–CO bonds, and inter-exchange of terminal carbonyls is easier than edge-bridging carbonyls with approximately similar Rh–CO bond lengths; this provides the first evidence for carbonyl exchange in carbonyl clusters occurring by a dissociative mechanism. Stereospecific CO-exchange in other carbonyl clusters has previously been observed, but these cases involve either the conversion of three-electron donor ligands to one-electron donor ligands (Bhaduri *et al.* 1979; Bryan *et al.* 1977) or the formation of multiply metal–metal bonded species (Johnson *et al.* 1978) with concomitant CO substitution by associative mechanism. Thus mechanisms of carbonyl inter-exchange can range from dissociative to associative and become further complicated if carbonyl intra-exchange (fluxionality) occurs.

(b) Carbonyl intra-exchange

This is extremely dependent on the geometry of the cluster and, for heterometallic clusters, the nature of the metals. Carbonyl migrations, which are still difficult to predict in advance, occur by either (a) the interconversion of terminal and bridging carbonyls, or (b) localized rocking or rotations of particular groups on a metal atom.

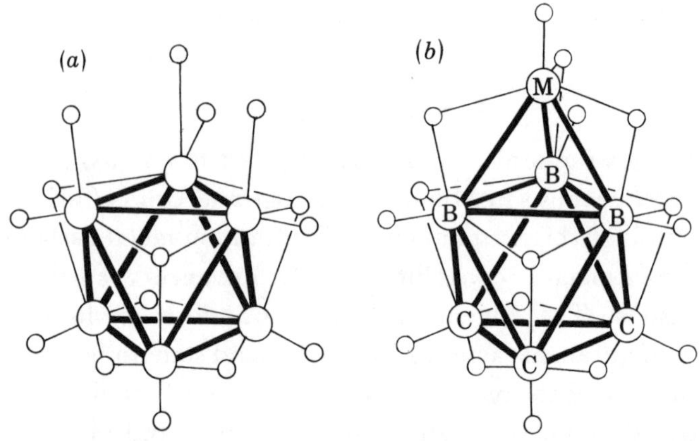

FIGURE 1. Schematic representation of the structures of (a) $[Rh_6(CO)_{15}]^{2-}$ and (b) $[MRh_6(CO)_{16}]^{n-}$ (M = Rh, $n = 3$; M = Ni, $n = 2$), which shows how little reorganization of the carbonyl skeleton occurs on capping a Rh_3-face.

We have found no evidence for the proposal that carbonyl fluxionality can be accounted for by the migration of a metal polyhedron within a fixed carbonyl polyhedron (Benfield & Johnson 1978).

The predictive difficulties in this area are illustrated by the gross change in carbonyl fluxionality found recently for the two iso-electronic, iso-structural clusters, $[Rh_7(CO)_{16}]^{3-}$ and $[NiRh_6(CO)_{16}]^{2-}$ (figure 1b).

Variable-temperature $^{13}C-\{^{103}Rh\}$ n.m.r. studies show that the only carbonyl migration in $[Rh_7(CO)_{16}]^{3-}$ involves the interchange of terminal and edge-bridging carbonyls around the Rh_3-triangular face remote from the capping atom (Brown *et al.* 1979a). However, $[NiRh_6(CO)_{16}]^{2-}$ exhibits three distinct carbonyl migrations; the lowest-energy migration involves a localized interchange of terminal and bridging carbonyls on Rh_B, followed by the same migration that was found in $[Rh_7(CO)_{16}]^{3-}$, followed by a complete carbonyl scrambling over the intact Ni-capped Rh_6-octahedron. It should be noted that the n.m.r. parameters ($\delta(^{13}CO)$, $\delta(^{103}Rh)$, $^1J(^{103}Rh-^{13}CO)$) for these two compounds are very similar except that the chemical shifts of the bridging carbonyls in $[NiRh_6(CO)_{16}]^{2-}$ occur at significantly higher field than in $[Rh_7(CO)_{16}]^{3-}$, which is consistent with the increased charge, on going from the dianion to the trianion, being mainly dissipated onto the bridging carbonyls (Heaton *et al.* 1982a).

It should also be noted that CO migrations can occur by the pairwise interconversion of terminal and bridging carbonyls, and this is facilitated if each step of the cyclic migration regenerates a structure identical to the original, as found in $[Rh_{13}(CO)_{24}H_{5-n}]^{n-}$ ($n = 1,2,3,4$) (Martinengo *et al.* 1977) and $[Rh_6(CO)_{13}C]^{2-}$ (Heaton *et al.* 1981*b*).

Isoelectronic clusters do not always adopt identical structures, as exemplified by the different structures of $[Rh_5(CO)_{15}]^-$ (Fumagalli *et al.* 1980) and $[FeRh_4(CO)_{15}]^{2-}$, which has been shown by n.m.r. measurements to be similar to $[RuIr_4(CO)_{15}]^-$ (Fumagalli *et al.* 1981). The subtle interplay of steric and electronic effects, which are currently not well understood, are responsible for the different structures of $[Rh_5(CO)_{15}]^-$ and $[FeRh_4(CO)_{15}]^{2-}$, and this in turn leads to different carbonyl migrational behaviour in solution (Ceriotti *et al.* 1982). Direct ^{103}Rh n.m.r. measurements provide no evidence for rearrangement of the apical iron in the FeRh$_4$-trigonal bipyramidal metal skeleton, whereas platinum is found to occupy both equatorial and apical positions in $[PtRh_4(CO)_{14}]^{2-}$ and $[PtRh_4(CO)_{12}]^{2-}$ respectively (Fumagalli *et al.* 1980*a*).

(c) *Intramolecular metal polyhedral rearrangements*

These were not found in the above experiments but there is an increasing number of examples of such rearrangements. The first example involved rotation of intact $Pt_3(CO)_6$ groups with respect to each other in the Pt$_3$-triangular stacked clusters, $[Pt_{3n}(CO)_{6n}]^{2-}$ ($n = 2,3,4$) (Brown *et al.* 1979*b*) and further examples involving the mono- and bi-capped square antiprismatic clusters $[Rh_9E(CO)_{21}]^{2-}$ (E = P, As), and $[Rh_{10}E(CO)_{22}]^{n-}$ (E = P, As, $n = 3$; E = S, $n = 2$) have now been found (Gansow *et al.* 1980; Garlaschelli *et al.* 1982).

(d) H *migration*

This can occur either around the periphery of the metallic skeleton as found for $[Ru_4(CO)_{11}\{P(OMe)_3\}H_4]$ (Knox *et al.* 1975) or within the metallic skeleton as found for $[Rh_{13}(CO)_{24}H_{5-n}]^{n-}$ ($n = 1,2,3,4$) (Martinengo *et al.* 1977) and $[Rh_{14}(CO)_{25}H]^{3-}$ (Heaton *et al.* 1980). The latter two examples resemble H diffusion in metals. At low temperatures, ($-90\ °C$), it is possible to stop H migration in $[Rh_{13}(CO)_{24}H_3]^{2-}$ and $^1H-\{^{103}Rh\}$ n.m.r. measurements show that the hydrides occupy the pseudo-octahedral holes that contain one and two edge-bridging carbonyls on the Rh$_4$-square faces of the hexagonal close-packed metal skeleton (figure 2*a*).

CLUSTER AGGLOMERATION BY CAPPING TRIANGULAR SQUARE FACES OF METAL POLYHEDRA

Capping of triangular or square-faces of metal polyhedra provides a controlled method for cluster growth and somewhat resembles surface reconstruction on metals.

The triangular faces of the trigonal prismatic cluster $[Rh_6(CO)_{15}C]^{2-}$ (Albano *et al.* 1973) can be capped, and n.m.r. measurements show that successive additions of Ag[BF$_4$] to this cluster result in the progressive formation of

$$[\{Rh_6(CO)_{15}C\}Ag\{Rh_6(CO)_{15}C\}]^{3-}, \ [\{Rh_6(CO)_{15}C\}Ag\{Rh_6(CO)_{15}C\}Ag\{Rh_6(CO)_{15}C\}]^{4-},$$
$$[\{Rh_6(CO)_{15}C\}Ag]^{n-}_n, \ [Ag\{Rh_6(CO)_{15}C\}Ag\{Rh_6(CO)_{15}C\}Ag\{Rh_6(CO)_{15}C\}Ag]^{2-},$$
$$[Ag\{Rh_6(CO)_{15}C\}Ag\{Rh_6(CO)_{15}C\}Ag]^-, \ \text{and} \ [Ag\{Rh_6(CO)_{15}C\}Ag].$$

The first-formed double-decker sandwich compound has been isolated and shown by X-ray crystallography to have staggered Rh_6-trigonal prismatic units bridged by Ag^I (V. G. Albano, personal communication 1982). It seems as though there is a pair of electrons located on the Rh_3-face in $[Rh_6(CO)_{15}C]^{2-}$, which is thus acting as a sophisticated ligand to Ag^I. Protonation of this face also occurs on the formation of $[HRh_6(CO)_{15}C]^-$ (Heaton $et\ al.$ 1982c), but in this case there is no evidence for H-bridged oligomers.

Related reactions on Rh_3-faces can also occur with nucleophilic reagents:

$$[Rh_6(CO)_{15}]^{2-} + [Rh(CO)_4]^- \underset{-70\,°C}{\overset{+25\,°C}{\rightleftharpoons}} [Rh_7(CO)_{16}]^{3-} + 3CO. \tag{1}$$

This reaction has been shown by ^{13}C n.m.r. to be completely reversible in MeOH under 1 atm CO, with higher temperatures favouring the larger cluster. This result appears to be quite general and there is obviously a delicate balance between cluster fragmentation and agglomeration on varying the temperature and pressure of CO.

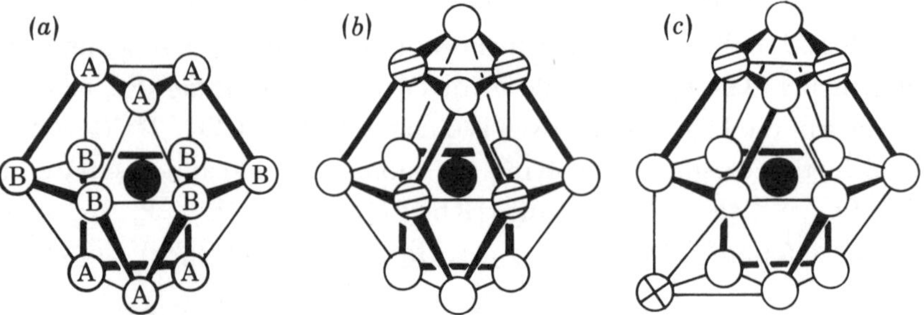

FIGURE 2. Schematic representation of the structures of (a) $[Rh_{13}(CO_{24}H_{5-n}]^{n-}$ ($n = 2,3,4$), (b) $[Rh_{14}(CO)_{25}H_{4-n}]^{n-}$ ($n = 3,4$) and (c) $[Rh_{15}(CO)_{27}]^{3-}$, which shows how little reorganization of the carbonyl skeleton occurs on capping a Rh_4-square-face. ●, Interstitial Rh_C; ○, rhodium with one terminal CO; ⊕, rhodium with two terminal CO's; ⊘, rhodium with no terminal CO's.

It should be noted that there is minimum reorganization of the carbonyl polyhedron on going from $[Rh_6(CO)_{15}]^{2-}$ to $[Rh_7(CO)_{16}]^{3-}$ (figure 1).

Electrophilic additions to square-faces occur in (2) and (3):

$$[Rh_{13}(CO)_{24}H]^{4-} + [Rh(CO)_2(CH_3CN)_2]^+ \longrightarrow [Rh_{14}(CO)_{25}H]^{3-} + CO; \tag{2}$$

$$[Rh_{14}(CO)_{25}]^{4-} + [Rh(CO)_2(CH_3CN)_2]^+ \longrightarrow [Rh_{15}(CO)_{27}]^{3-}. \tag{3}$$

Again it should be noted that minimum reorganization of the carbonyl polyhedron occurs (figure 2). The two square-faces that are capped are the same as two of the three faces that are occupied by hydrogen at low temperature in $[Rh_{13}(CO)_{24}H_3]^{2-}$, and this probably results because these faces are situated beneath large pentagonal holes on the periphery of the carbonyl polyhedron in $[Rh_{13}(CO)_{24}H_{5-n}]^{n-}$ ($n = 1,2,3$) (Ciani $et\ al.$ 1981; Albano $et\ al.$ 1975, 1979).

BONDING WITHIN THE CLUSTER

(a) Carbonyls

Charge delocalization is most effectively accomplished by bridging carbonyls, and for isoelectronic clusters there is a significant shift of $\delta(^{13}CO)_{mean}$ to lower field with increasing charge on the anion. Recent CNDO calculations (Freund & Hohlneicher 1979) on $[Co_4(CO)_{12}]$ showed that the axial carbonyl carbon is most positive and thus involved in the least metal $\rightarrow$ carbonyl back-bonding. This is also consistent with the axial carbonyl ^{13}C resonance in $[Rh_4(CO)_{12}]$ derivatives always occurring at highest field (Heaton *et al.* 1982b) and with their patterns of substitution. The first ligand always occupies an axial site in $[M_4(CO)_{12}]$ (M = Co, Rh, Ir) and then depending on steric effects as to whether the next ligand also goes into the electronically preferred but now sterically hindered axial site or into a radial site.

TABLE 1. ROOM-TEMPERATURE N.M.R. DATA† (SEE FIGURE 2) FOR $[Rh_{13}(CO)_{24}H_{5-n}]^{n-}$
(n = 1, 2, 3, 4)

n =	4	3	2	1
$10^6\delta(^{103}Rh_A)\ddagger$	-340	-403	-532	-527
$10^6\delta(^{103}Rh_B)\ddagger$	-240	-522	-600	-536
$10^6\delta(^{103}Rh_C)\ddagger$	$+6370$	$+4554$	$+3547$	$+2917$
$10^6\delta(^1H)\ddagger$	-25.5	-26.7	-29.3	-31.2
$J(Rh_A-H)/Hz$	5.2	4.4	4.8	4.9
$J(Rh_B-H)/Hz$	5.2	5.8	5.7	4.9
$J(Rh_C-H)/Hz$	17.1	22.3	23.2	18.6

† $\delta(^{103}Rh)\ 0 \times 10^{-6}$ = 3.16 MHz at such a magnetic field that the protons in TMS resonate at exactly 100 MHz.
‡ The factor 10^{-6} is used to denote the notion of parts per million.

(b) Metals

Interstitial rhodium atom resonances are always at high frequency in a region normally associated with positive oxidation states. For $[Rh_{13}(CO)_{24}H_{5-n}]^{n-}$ (n = 1, 2, 3, 4), increasing the number of hydrides migrating interstitially around the interstitial rhodium atom, Rh_C, results in a dramatic shift of $\delta(^{103}Rh_C)$ to lower frequency while the peripheral rhodium atoms, Rh_A and Rh_B, vary very little (table 1). This is consistent with increased diamagnetic shielding of Rh_C and suggests that interstitial metal atoms are more positive than surface metal atoms, which is in keeping with bulk metallic properties. It seems probable that the large variation in $\delta(^{103}Rh)$ that alternates along the C_3-axis in $[Rh_7(CO)_{16}]^{3-}$ and along the C_4-axis is $[Rh_{14}(CO)_{25}H_{4-x}]^{x-}$ (x = 3, 4) (Brown *et al.* 1979a; Heaton *et al.* 1980) could also be ascribed to alternation of charge along the principal axis of the metallic skeleton, with capping atoms exhibiting electropositive character similar to kink and step sites on metallic surfaces.

(c) Main group elements in interstitial sites

These have already been discussed (Chini 1980) and the recent low-field shift of the carbide resonance on going from the trigonal prismatic cluster, $[Rh_6(CO)_{15}C]^{2-}$, to the octahedral cluster, $[Rh_6(CO)_{13}C]^{2-}$, is in keeping with a contraction of the radius of carbon and an increase in positive charge (Heaton *et al.* 1981).

Recent infrared work on the isostructural, isoelectronic clusters $[M_6(CO)_{15}C]^{2-}$ and $[M_6(CO)_{15}N]^-$ (M = Co and Rh) showed the metal–interstitial atom force constants to be very similar for the M_6–C and M_6–N groups with the values for M = Rh being significantly smaller than when M = Co (Creighton *et al.* 1982).

N.M.R. MEASUREMENTS UNDER HIGH PRESSURES OF GAS

Homogeneous catalytic reactions often require high pressures and elevated temperatures, and until recently high-pressure infrared spectroscopy has been the only spectroscopic technique available to monitor such reactions. However, n.m.r. probes capable of measurements under high pressures of gas (up to 1 kbar (10^5 Pa)) have recently been developed and these will provide more unambiguous spectroscopic evidence about catalytic intermediates and mechanisms than has been available hitherto. The first studies in this area were connected with the catalytic synthesis of ethylene glycol from synthesis gas as it was shown that $[Rh_5(CO)_{15}]^-$ was the only species present under 1 kbar CO–H_2 at 25 °C (Heaton *et al.* 1981 *a*).

Further developments are in hand to allow measurements at higher temperatures so that *in situ* measurements under catalytic conditions can be made. This technique promises to be extremely important, especially as it can readily be extended to other nuclei (e.g. ^{1}H, ^{13}C, ^{31}P).

I thank the S.E.R.C. for their financial support of this work, the preparative expertise of Dr L. Strona and Dr R. Della Pergola in Canterbury and the group of cluster chemists in Milan, Dr R. J. Goodfellow for ^{1}H–{^{103}Rh} measurements, Dr I. H. Sadler for ^{103}Rh measurements, and Professor J. Jonas for the use of his spectrometer for the initial high-pressure n.m.r. experiments.

References

Albano, V. G., Ceriotti, A., Chini, P., Ciani, G., Martinengo, S. & Anker, W. M. 1975 *J. chem. Soc. chem. Commun.*, pp. 859–860.

Albano, V. G., Ciani, G., Martinengo, S. & Sironi, A. 1979 *J. chem. Soc. Dalton Trans.*, pp. 978–982.

Albano, V. G., Sansoni, M., Chini, P. & Martingneo, S. 1973 *J. chem. Soc. Dalton Trans.*, pp. 651–655.

Benfield, R. E. & Johnson, B. F. G. 1978 *J. chem. Soc. Dalton Trans.*, pp. 1554–1568.

Bhaduri, S., Johnson, B. F. G., Lewis, J., Watson, D. J. & Zuccaro, C. 1979 *J. chem. Soc. Dalton Trans.*, pp. 557–561.

Brown, C., Heaton, B. T., Longhetti, L., Smith, D. O., Chini, P. & Martinengo, S. 1979 *a J. organometall. Chem.* **169**, 309–314.

Brown, C., Heaton, B. T., Towl, A. D. C., Chini, P., Fumagalli, A. & Longoni, G. 1979 *b J. organometall. Chem.* **181**, 233–254.

Bryan, E. G. Johnson, B. F. G. & Lewis, J. 1977 *J. chem. Soc. chem. Commun.*, pp. 329–330.

Ceriotti, A., Longoni, G., Manassero, M., Sansoni, M., Della Pergola, R., Heaton, B. T. & Smith, D. O. 1982 *J. chem. Soc. chem. Commun.*, pp. 886–887.

Chini, P. 1980 *J. organometall. Chem.* **200**, 37–61.

Ciani, G., Sironi, A. & Martinengo, S. 1981 *J. Chem. Soc. Dalton Trans.*, pp. 519–523.

Creighton, J. A., Della Pergola, R., Heaton, B. T., Martinengo, S., Strona, L. & Willis, D. 1982 *J. chem. Soc. chem. Commun.*, pp. 864–865.

Freund, H. J. & Hohlneicher, G. 1979 *Theor. chim. Acta* **51**, 145–162.

Fumagalli, A., Koetzle, T. F. & Takusagawa, F. 1981 *J. organometall. Chem.* **213**, 365–377.

Fumagalli, A. Koetzle, T. F., Takusagawa, F., Chini, P., Martinengo, S. & Heaton, B. T. 1980 *b J. Am. chem. Soc.* **102**, 1740–1742.

Fumagalli, A., Martinengo, S., Chini, P., Albinati, A. & Bruckner, S. 1980 *a* In *XIII Meeting of the Italian Chemical Society, Camerino*, A 11.

Gansow, O. A., Gill, D. S., Bennis, F. J., Hutchinson, J. R., Vidal, J. L. & Schoening, R. C. 1980 *J. Am. chem. Soc.* **102**, 2449–2456.

Garlaschelli, L., Fumagalli, A., Martinengo, S., Heaton, B. T., Smith, D. O. & Strona, L. 1982 *J. Chem. Soc. Dalton Trans.* (Submitted.)

Heaton, B. T., Brown, C., Smith, D. O., Strona, L., Goodfellow, R. J., Chini, P. & Martinengo, S. 1980 *J. Am. chem. Soc.* **102**, 6177–6178.

Heaton, B. T., Della Pergola, R. Strona, L., Smith, D. & Fumagalli, A. 1982 *a J. chem. Soc. Dalton Trans.* (Submitted.)

Heaton, B. T., Jonas, J., Eguchi, T. & Hoffman, G. A. 1981 *a J. chem. Soc. chem. Commun.*, pp. 331–332.

Heaton, B. T., Strona, L., Della Pergola, R., Garlaschelli, L., Sartorelli, U. & Sadler, I. H. 1982 *b J. chem. Soc. Dalton Trans.* (Submitted.)

Heaton, B. T., Strona, L. & Martinengo, S. 1981 *b J. organometall. Chem.* **215**, 415–422.

Heaton, B. T., Strona, L., Martinengo, S., Strumolo, D., Goodfellow, R. J. & Sadler, I. H. 1982 *c J. chem. Soc. Dalton Trans.*, pp. 1499–1502.

Johnson, B. F. G., Lewis, J. & Pippard, D. 1978 *J. organometall. Chem.* **160**, 263–274.

Knox, S. A. R., Keopke, J. W., Andrews, M. A. & Kaesz, H. D. 1975 *J. am. chem. Soc.* **94**, 3942–3952.

Martinengo, S., Heaton, B. T., Goodfellow, R. J. & Chini, P. 1977 *J. chem. Soc. chem. Commun.*, pp. 39–40.

Discussion

JOAN MASON (*Chemistry Department, The Open University, U.K.*). Ed Randall has mentioned the usefulness of the quadrupolar nucleus ^{17}O in n.m.r. studies of metal carbonyl clusters. We found that both ^{14}N and ^{15}N n.m.r. spectroscopy were interesting and informative in the study of interstitial nitrogen in the trigonal prismatic clusters of cobalt and rhodium, $[PPh_4][M_6N(CO)_{15}]$ (Martinengo *et al.* 1979). This is because the quadrupolar ^{14}N nucleus gives relatively sharp lines when the local symmetry is high, as well as the line width containing information on this symmetry. Indeed, the line widths were similar ($W_{\frac{1}{2}} \sim 40$ Hz) in ^{14}N and in ^{15}N resonance (with ^{15}N enrichment) in the cobalt cluster, with line broadening due to the ^{59}Co quadrupole and to unresolved cobalt–nitrogen coupling. The ^{14}N and ^{15}N lines were sharper for the rhodium cluster, because ^{103}Rh has $I = \frac{1}{2}$. The ^{14}N line width was now 20 Hz, no spin–spin coupling being resolved; but ^{15}N resonance, with enrichment, yielded the central five lines of the expected septet, with $J_{Rh^{15}N} = 6$ Hz and $W_{\frac{1}{2}} = 1$ Hz. This coupling constant matches the $J_{Rh^{13}C} = 14$ Hz observed for interstitial carbide in the isostructural complexion $[Rh_6C(CO)_{15}]^{2-}$ (given the different magnetogyric ratios), which suggests that the charge dispositions may be similar for the nitride and carbide.

The rhodium–^{15}N coupling constant of 6 Hz corresponds to a ^{14}N value of 4 Hz, so the ^{14}N line width in the absence of unresolved coupling may be as small as 5–8 Hz. Coupling to interstitial ^{14}N might therefore be measurable for metal nuclei with higher magnetogyric ratios.

Comparison of the ^{14}N and ^{15}N line widths indicates broadening by quadrupolar relaxation for ^{14}N; the local symmetry is less than spherical, and molecular vibrations and collisions give rise to transient electric field gradients. Smaller line widths, less than 1 Hz, are observed for aqueous $^{14}NH_4^+$ and $^{14}NMe_4^+$, and these are explained in part by shorter correlation times τ_q for the smaller ions, because $W_{\frac{1}{2}}$ is proportional to τ_q. The observed quadrupolar relaxation time $T_q (= 1/\pi W_{\frac{1}{2}} \approx 50$ ms) for nitride in the rhodium cluster, with a correlation time of say, 10^{-10} s, corresponds to a nuclear electric quadrupole coupling constant of about 1 MHz. This is larger than has been observed for NH_4^+ in solids (Wolff *et al.* 1977) and so may correspond to slightly lower symmetry for nitrogen in the cluster.

The nitrogen shifts were unexpected. We observed (Martinengo *et al.* 1979) the customary increase in shielding down the group of the transition metal (from cobalt to rhodium) for the nitrides, as for the isostructural carbides in ^{13}C resonance; but the lines were at relatively high field in nitrogen resonance and at low field in carbon resonance. The radial factor $\langle r^{-3} \rangle_{2p}$ in the paramagnetic shielding term commonly gives nitrogen shifts (measured downfield of NH_4^+) that are roughly double the carbon shifts (measured downfield of CH_4) in isoelectronic molecules, the deshielding circulations of the 2p electrons being closer to the more electronegative nucleus. The reversal that we observe for the isoelectronic carbide and nitride points to the dominant factor being the energy denominator (in the paramagnetic term) which represents the h.o.m.o.–l.u.m.o. gap for the rotation of electronic charge in the magnetic field (Mason 1981). Certainly the range of chemical shifts observed for interstitial atoms in (diamagnetic) metal carbonyl clusters, which is 54×10^{-6} for ^1H so far, is larger than could be accounted for by differences in atomic charge (which act upon the radial factor). X-ray crystallographic measurements give the carbon and nitrogen radii within the Co_6 trigonal prism as 74 and 67 pm respectively (Martinengo *et al.* 1979), so perhaps there is greater overlap and splitting of the relevant orbitals in the nitride, increasing the h.o.m.o.–l.u.m.o. gap for charge rotation. (Interestingly,

$[Co_6C(CO)_{15}]^{2-}$ is black and $[Co_6N(CO)_{15}]^-$ orange-red!) There is, however, no simple relation between ^{13}C shift and cavity size for the small number of interstitial carbides so far reported; as we might expect, the cluster symmetry is also important.

References

Martinengo, S., Ciani, G., Sironi, A., Heaton, B. T. & Mason, J. 1979 *J. Am. chem. Soc.* **101**, 7095.
Mason, J. 1981 *Chem. Rev.*, p. 205.
Wolff, E. K., Griffin, R. G. & Watson, C., 1977 *J. chem. Phys.* **66**, 5433.

B. T. HEATON. The supposition by Dr Mason that comparison of $^1J(^{103}Rh-^{13}C)$ and $^1J(^{103}Rh-^{15}N)$ in $[Rh_6(CO)_{15}E]^{n-}$ (E = C, $n = 2$; E = N, $n = 1$) provides information on charge disposition on the interstitial atom is erroneous. While agreeing with her final statement on the importance of cluster symmetry in determining the interstitial atom chemical shift, it seems premature to state that 'there is no simple relation between ^{13}C shift and cavity size' because, apart from Bradley's work (see his paper in this symposium), there is only one other report (Heaton *et al.* 1981 *b*) of interstitial atom n.m.r. data on closely related clusters containing the same number of the same metal atoms. Comparisons outside such closely related clusters are positively hazardous at the moment!

E. W. RANDALL (*Chemistry Department, Queen Mary College, London, U.K.*). Since Dr Heaton used material enriched in ^{13}C for his n.m.r. studies did he consider measuring ^{13}C T_1 values?

B. T. HEATON. Few reports have currently appeared on ^{13}C T_1 values of carbonyl clusters. It would be very interesting to obtain this information but so far we have been prevented from carrying out these experiments by the limited spectrometer time available to me.

E. W. RANDALL. One advantage that ^{17}O has over ^{13}C is the generally fast relaxation (up to 1000 times faster than for ^{13}C), which allows relatively easy measurement of T_1, $T_{1\rho}$ and T_2. These quantities yield information on the overall dynamics of the cluster, including fluxional processes, and a comparison of these quantities for any one centre enables a separation of the different processes to be made. Additionally, in non-fluxional limits which yield resolved environments it is possible in the relatively isotropic overall tumbling case of large clusters to deduce from differences in T_1 values for different sites quite pronounced differences in the contribution of the term involving the nuclear electric quadropole to the ^{17}O relaxation. Dr G. E. Hawkes and Dr S. Aime have found a factor of about 500 in this (quadrupolar) term for the two CO environments of $Rh_6(CO)_{16}$.

Phil. Trans. R. Soc. Lond. A **308**, 103–113 (1982)   [103]

Printed in Great Britain

Carbidocarbonyl clusters of iron

By J. S. Bradley

Corporate Research Laboratories, Exxon Research and Engineering Company,
P.O. Box 45, Linden, New Jersey 07036, U.S.A.

Under reducing conditions, at elevated temperatures, coordinated carbon monoxide in transition metal carbonyls may disproportionate to CO_2 and a carbon atom. The carbon atom is trapped in a cage of metal atoms, shielded from further reaction in the core of the resulting carbidocarbonyl clusters. This class of compounds, which has been known for some time, bears some structural resemblance to the binary carbides and may therefore be relevant to Fischer–Tropsch catalysis, in which carbide phases and surface carbon atoms are implicated.

The chemical, structural and physical properties of the iron carbidocarbonyls have been investigated. Reactions are described that lead to C–H and C–C bond formation at the carbide carbon, and these are discussed with regard to the nature of this carbon atom. ^{13}C n.m.r. spectroscopy reveals large downfield chemical shifts for the carbide carbon, which may be interpreted as either a reflexion of low electron density at the carbon or the influence of paramagnetic contributions to the shift. Reactions of $[Fe_4(CO)_{12}C\cdot CO_2CH_3]^-$, $[Fe_4(CO)_{12}C\cdot C(O)CH_3]^-$ and $[Fe_4(CO)_{12}C\cdot CHO]^-$ with trimethyloxonium fluoborate yield the corresponding vinylidene clusters $Fe_4(CO)_{12}$-$C{=}C(OCH_3)R$ $(R = OCH_3, CH_3, H)$.

1. Introduction

In 1962 Dahl and coworkers reported the synthesis and structure of an iron carbonyl cluster with a unique structural feature (Braye *et al.* 1962). $[Fe_5C(CO)_{15}]$, the new molecule contained, in addition to its carbonyl ligands, a single carbon atom bound only to the five iron atoms in the molecule. This first example of a carbidocarbonyl cluster remained the only example of its type for several years, until the isolation by Lewis and coworkers of $[Ru_6C(CO)_{17}]$ and its derivatives (Johnson *et al.* 1967, 1968). In the mid 1970s, the late Paolo Chini and his coworkers began reporting their successes in the synthesis and structural characterization of a number of cobalt and rhodium carbidocarbonyl clusters, and a similar acceleration occurred in the rate of growth of the analogous chemistry of iron, ruthenium and osmium. As the array of carbido-carbonyl clusters has grown, so their significance as a class has become more obvious. What were once regarded as structural oddities have assumed a more central role in cluster chemistry, at a time when the ostensible similarities between metal cluster compounds and metallic crystallites were being explored in terms of both structure and reactivity (Ugo 1975; Muetterties 1977).

An additional significance for carbidocarbonyl clusters has appeared in the past 3 years, with the discovery of the fascinating reactivity of carbon atoms in low nuclearity iron clusters when they are exposed to reactive molecules. These observations followed on the heels of the recognition of the crucial role played by surface-bound carbon atoms in metal-catalysed carbon monoxide hydrogenation (Ponec 1978), and so a new area of overlap between cluster chemistry and surface chemistry has arisen. Moreover, in this case the comparisons between organometallic and surface chemistry may lie not only in structural similarities but also in chemical reactivity. In this paper, I shall describe some recent developments in the chemistry of the carbidocarbonyl clusters of iron, with emphasis on the reactivity of the cluster-bound carbon atom.

2. Tetranuclear carbidocarbonyl clusters of iron

As mentioned above, the first carbidocarbonyl transition metal cluster to be recognized was $[Fe_5C(CO)_{15}]$, **1**, isolated in very low yield from the reaction of triiron dodecacarbonyl with methylphenylacetylene and characterized by X-ray diffraction by Dahl and coworkers (Braye *et al.* 1962). The molecule comprises a square pyramidal Fe_5 core with the carbide carbon situated 0.08 Å† below the centre of the square face. The hexanuclear analogue to **1** is known, as a dianion. $[Fe_6C(CO)_{16}]^{2-}$, **2**, originally synthesized by the reaction of $Fe(CO)_5$ with a variety of metal carbonyl anions at elevated temperatures, contains an octahedral Fe_6C core, with the carbon atom situated at its centre (Churchill *et al.* 1971).

The defining characteristic of this class of compounds is the lone carbon atom, bound only to metal atoms. Although the metal-centred chemistry of these clusters is undoubtedly affected by the carbon atom, these effects (for example in ligand exchange at the metal) are secondary when compared with the potential for unique chemistry at this carbon, caused by its surrounding metal atoms. However, in neither **1** nor **2** has any reactivity been observed at the carbon atom. This is hardly surprising for the fully encapsulated carbon in **2**, but less expected for **1**, in which the carbon atom is at least partly exposed (Tachikawa & Muetterties 1981).

Iron is so far unique in that it alone forms carbidocarbonyl clusters containing only four metal atoms, and the discovery of this class of compounds and of the reactivity of its members has provided one of the more intriguing developments in cluster chemistry over the last few years. We reported in 1979 that the cluster-bound carbon atom, which was apparently inert in Fe_5C and Fe_6C clusters, could display novel chemistry if exposed by the removal of some of its surrounding metal atoms, reducing the coordination number of the carbon atom to four in an open, butterfly geometry (Bradley *et al.* 1979). This, the first observation of the reactivity of a cluster-bound carbon atom, resulted from the partial fragmentation of the octahedral carbide cluster $[Fe_6C(CO)_{16}]^{2-}$, **2**. When treated with tropylium bromide, a mild one-electron oxidizing agent, in methanol, **2** loses two iron vertices (as ferrous ion), exposing the central carbon atom, which reacts with CO and methanol yielding the μ^4-carbomethoxymethylidyne cluster, **3**:

$$[Fe_6C(CO)_{16}]^{2-} \xrightarrow[\text{CH}_3\text{OH, 25 °C}]{\text{C}_7\text{H}_7^+\text{Br}^-} [Fe_4(CO)_{12}C\cdot CO_2CH_3]^- + 2Fe^{2+} + 3CO. \qquad (1)$$
$$\underset{\textbf{2}}{\phantom{[Fe_6C(CO)_{16}]^{2-}}} \qquad \underset{\textbf{3}}{\phantom{[Fe_4(CO)_{12}C\cdot CO_2CH_3]^-}}$$

The structure of **3** (figure 1), comprises an open, butterfly arrangement of four iron atoms, each bearing three terminal carbonyls. The hitherto encapsulated carbon atom is equidistant (2.00 ± 0.02 Å) from the metal atoms and is now bonded to the carbomethoxy group. The nature of this reaction will be discussed below in the context of other carbon–carbon bond-forming reactions.

The reactivity of the carbon atom when bound to an open Fe_4 cluster has an obvious relevance to the chemistry of surface-bound carbon atoms in heterogeneous catalysis, and is of great interest. Several members of the Fe_4C family have been synthesized and structurally characterized, and the chemistry of these clusters is under investigation in several laboratories.

† 1 Å $= 10^{-10}$ m $= 0.1$ nm.

The parent neutral carbido-cluster of this class, $[Fe_4C(CO)_{13}]$, **4**, is most readily synthesized by protonation of **3** (Bradley *et al.* 1981):

$$[Fe_4(CO)_{12}C \cdot CO_2CH_3]^- \xrightarrow[\text{methylcyclohexane}]{CF_3SO_3H} [Fe_4C(CO)_{13}]. \qquad (2)$$
$$\textbf{3}\textbf{4}$$

The molecule (figure 2) comprises an open, butterfly-shaped Fe_4C core, with three terminal carbonyls bound to each ion atom. The thirteenth carbonyl bridges the two iron atoms, which make up the central bond $(Fe(1)-Fe(4))$ of the butterfly. Further aspects of the structure are described below in comparison with other members of this family.

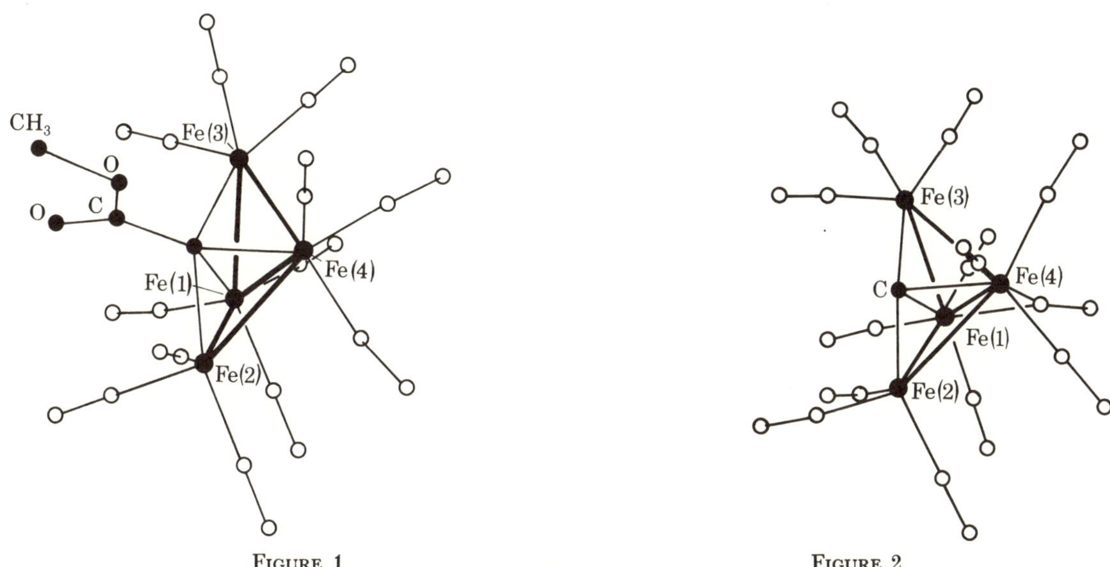

FIGURE 1   FIGURE 2

FIGURE 1. $Fe_4(CO)_{12}C \cdot CO_2CH_3^-$, **3**, as in its $(C_2H_5)_4N^+$ salt. A butterfly core of four iron atoms bears twelve terminal carbonyls, three on each iron. $Fe(1)-Fe(4)$ $(2.590(1)$ Å$)$ is significantly longer than the other Fe–Fe bonds $(2.503(9)$ Å mean$)$. The methylidyne carbon is a $2.027(9)$ Å from $Fe(2)$, $2.012(9)$ Å from $Fe(3)$, $1.957(9)$ Å from $Fe(1)$ and $1.952(9)$ Å from $Fe(4)$. $Fe(1)$, $Fe(4)$ and the carbomethoxy group are coplanar. The dihedral between $Fe(1)(3)(4)$ and $Fe(1)(2)(4)$ is $130°$. Angle $Fe(2)CFe(3) = 148°$.

FIGURE 2. $Fe_4C(CO)_{13}$, **4**. The butterfly Fe_4C core bears twelve terminal carbonyls and a thirteenth bridging $Fe(1)-$ $Fe(4)$. Unbridged Fe–Fe distances average $2.642(4)$ Å; $Fe(1)-Fe(4) = 2.545(1)$ Å, $Fe(1)-C = 1.998(4)$ Å, $Fe(4)-C = 1.987(4)$ Å, $Fe(2)-C = 1.797(4)$ Å; $Fe(3)-C = 1.800(4)$ Å. Dihedral angle between $Fe(1)(3)(4)$ and $Fe(1)(2)(4) = 101°$. Angle $Fe(2)CFe(3) = 175°$.

We have synthesized several phosphine-substituted derivatives of **4** by reaction of trimethylphosphine with **4**. $[Fe_4C(CO)_{13-n}(PMe_3)_n]$ $(n = 3$ or $4)$ have been fully characterized spectroscopically, and the structure of $[Fe_4C(CO)_{10}(PMe_3)_3]$ is shown in figure 3. The three phosphine ligands are bonded to three of the iron atoms, one wingtip iron at $Fe(3)$ remaining unsubstituted. A triply bridging carbonyl resides on the wing of the Fe_4C butterfly that contains $Fe(3)$. The tetrakis(trimethylphosphine) derivative $[Fe_4C(CO)_9(PMe_3)_4]$ has been characterized by ^1H, ^{31}P and ^{13}C n.m.r. spectroscopy, and on the basis of the presence of two equally populated phosphine sites is assigned the symmetric structure, each iron atom bearing one phosphine ligand and two carbonyls.

Reduction of **4** by sodium amalgam yields $[Fe_4C(CO)_{12}]^{2-}$, **5**, in high yield:

$$[Fe_4C(CO)_{13}] \xrightarrow[25\,°C]{Na-Hg, \, THF} [Fe_4C(CO)_{12}]^{2-}. \qquad (3)$$
$$\textbf{4}\textbf{5}$$

Compound **5** was previously synthesized by Tachikawa and Muetterties (Tachikawa *et al.* 1980), and structurally characterized as its $Zn(NH_3)_4^{2+}$ salt. We have determined independently the structure of the tetraethylammonium salt of **5** (in which the anion is essentially of the same structure as in the zinc salt), and this is shown in figure 4. The anion is similar in geometry to its neutral analogue **4** but lacking the bridging carbonyl found in the latter.

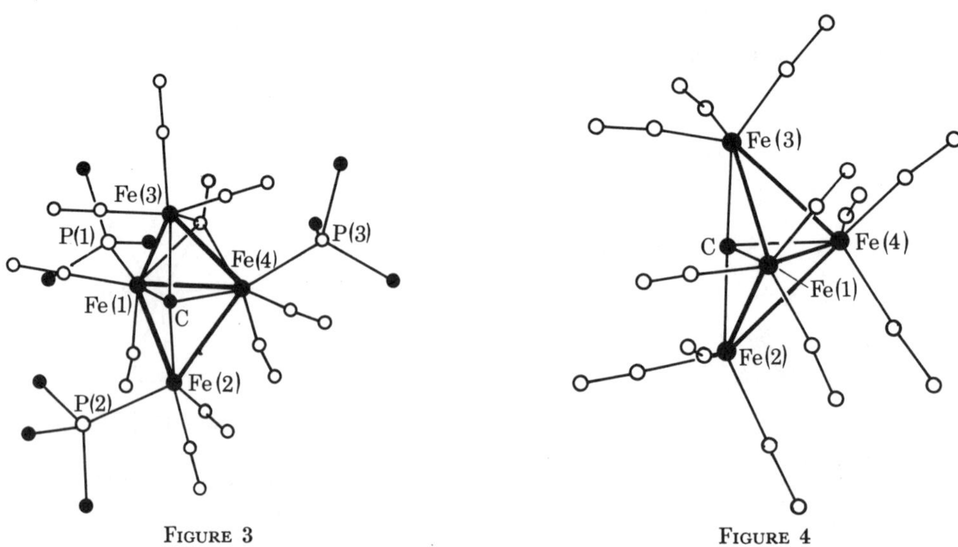

FIGURE 3　　　　　　　　　　　　FIGURE 4

FIGURE 3. $Fe_4C(CO)_{10}(P(CH_3)_3)_3$. The three trimethylphosphine ligands are coordinated to Fe(1), Fe(2) and Fe(4). Of the ten carbonyl ligands nine are terminal, and one is triply bridging on the Fe(1)(3)(4) face. Fe(1)–Fe(4) = 2.528(1) Å, the shortest of the metal–metal bonds. The other Fe–Fe distances average 2.644(19) Å. Fe(1)–C = 1.979(4) Å, Fe(4)–C = 2.014(4) Å, Fe(2)–C = 1.785(4) Å, Fe(3)–C = 1.838(4) Å. The dihedral angle between Fe(1)(2)(4) and Fe(1)(3)(4) = 102.4°. Angle Fe(2)CFe(3) = 174°.

FIGURE 4. $Fe_4C(CO)_{12}^{2-}$, **5**, as in its $(C_2H_5)_4N^+$ salt. Fe(1)–Fe(2) = Fe(1)–Fe(3) = 2.640(2) Å. Fe(2)–Fe(4) = Fe(3)–Fe(4) = 2.641(3) Å. Fe(1)–Fe(4) = 2.521(2) Å. Fe(1)–C = 1.936(12) Å, Fe(2)–C = 1.772(11) Å, Fe(3)–C = 1.803(12) Å, Fe(4)–C = 1.969(12) Å. Dihedral at Fe(1)–Fe(4) = 101°. Angle Fe(2)Fe(3) = 178°.

Oxidation of **5** by ferrocenium cation in methylene chloride under carbon monoxide regenerates **4**.

Reduction of $[Fe_4(CO)_{12}CCO_2CH_3]^-$, **3**, with $BH_3 \cdot THF$ in THF proceeds smoothly at 25 °C to give $[HFe_4C(CO)_{12}]^-$, **6**, isolated in high yield as the tetraethylammonium salt. This compound has been reported independently by Davis *et al.* (1981), and by Holt *et al.* (1981), who also determined the structure of the bis(triphenylphosphineiminium) salt, locating the hydride on the central Fe–Fe bond of the Fe_4C butterfly.

3. STRUCTURAL FEATURES OF Fe_4C CLUSTERS

The three isoelectronic clusters $[Fe_4C(CO)_{13}]$, **4**, $[Fe_4C(CO)_{12}]^{2-}$, **5**, and $[HFe_4C(CO)_{12}]^-$, **6**, provide the central characters in the Fe_4C family. Their structures have been determined by X-ray diffraction, and in all three cases there is a basic $[Fe_4C(CO)_{12}]$ core of relatively invariant geometry, with the open, butterfly arrangement characteristic of Fe_4C clusters. Consistent with the predictions of Wade's rules (Wade 1972), these 62-electron clusters, possessing seven skeletal electron pairs and four vertices, adopt an *arachno* structure based on the octahedron. The carbonyls in the $Fe_4C(CO)_{12}$ unit are all terminal in groups of three on each iron atom, and the carbido carbon atom is situated centrally above the midpoint of the central Fe–Fe bond, midway between the two 'wingtips'. The dihedral angle between the two triangular wings of the butterfly averages 102 ± 2 Å over the three clusters (table 1), the angle subtended at the carbide carbon atom by the wingtip iron atoms averages $176 \pm 2°$, and the $Fe–C_{carbide}$ distances a and b are almost constant (see figure 5).

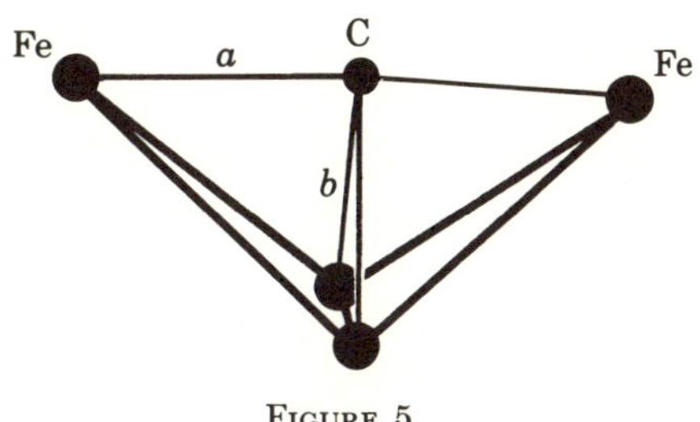

FIGURE 5

TABLE 1

	dihedral δ	angle FeCFe	a	b	
	deg	deg	Å	Å	references
$[Fe_4C(CO)_{13}]$, **4**	101	175	1.80	1.94	Bradley *et al.* (1981)
$[Fe_4C(CO)_{12}]^{2-}$, **5**	101	178	1.79	1.97	this work and Tachikawa *et al.* (1980)
$[Fe_4C(CO)_{12}H]^-$, **6**	104	174	1.79	1.99	Holt *et al.* (1981)

The dihedral angle δ is rather less than the 109° provided by a regular octahedral cavity (from which the butterfly is derived), and this compression is reflected in the disparity between the two sets of $Fe–C_{carbide}$ distances a and b (see figure 5).

It is also interesting in this respect to compare the geometry of the Fe_4C family with that of the hexanuclear and pentanuclear homologues $[Fe_6C(CO)_{16}]^{2-}$, and $[Fe_5C(CO)_{15}]$. If it were true that the octahedral cavity in the dianion was too small for the adequate accommodation of the carbon atom, it would be expected that in the more open Fe_5C and Fe_4C clusters some relaxation would occur, allowing the carbon atom to adopt a less crowded position. In fact the effective radius of the carbon atom in $[Fe_5C(CO)_{15}]$ is 0.57 Å (Braye *et al.* 1962), an increase of only 0.02 Å from that in $[Fe_6C(CO)_{16}]^{2-}$ (Churchill *et al.* 1971), and in the even more open structures of **4, 5** and **6** the average value is still only 0.58 Å. (These last values are derived by using a mean of the five Fe–Fe distances for each molecule.)

4. CHEMICAL REACTIVITY OF THE CARBON ATOM IN Fe_4C CLUSTERS

This aspect of the chemistry of carbidocarbonyl clusters is a relatively recently observed phenomenon. Carbon atom chemistry is found so far exclusively in the Fe_4C group of clusters, although increased activity in this area will undoubtedly uncover analogous reactions with other metals.

A number of reactions of Fe_4C clusters have been observed in which either C–C or C–H bonds to the carbide carbon are formed.

(a) C–C *bond formation*

(i) *Reaction of* $Fe_4C(CO)_{13}$ *with alcohols*

$$Fe_4C(CO)_{13} + ROH \longrightarrow Fe_4(CO)_{13}CCO_2R^- + H^+ \ (R = CH_3, C_2H_5, \text{iso-}C_3H_7). \quad (4)$$

The reaction, either in undiluted alcohol or in an inert polar solvent, is reversible on evacuation to remove solvent (Bradley *et al.* 1981). Addition of base is required to allow the isolation of the crystalline product, which has been fully characterized crystallographically for $R = CH_3$ (Bradley *et al.* 1979) as its tetraethyl ammonium salt (see figure 1). The *C*-carboalkoxy group is bound to the hitherto exposed carbon atom of the Fe_4C core, which has distorted from its original butterfly shape ($101°$ dihedral, angle $FeCFe = 175°$) to a bipyramidal form with the μ^4-methylidyne carbon atom at an equatorial vertex, consistent with the 60 valence electrons in the cluster (Wade 1972).

(ii) *Reaction of* $Fe_4C(CO)_{13}$ *with secondary amines*

$$\underset{\mathbf{4}}{Fe_4C(CO)_{13}} + 2R_2NH \longrightarrow (R_2NH_2)^+[Fe_4(CO)_{12}C \cdot C(O)NR_2]^-. \quad (5)$$

The addition of two equivalents of the amine to a methylene chloride solution of **4** at $25\,°C$ results in the rapid formation of the *C*-dialkylamide derivatives ($R = C_2H_5$, iso-C_3H_7). Although no X-ray structural data have been obtained on these compounds, their infrared and ^1H n.m.r. spectra are consistent with their formulation, and presumably they have structures similar to the carboalkoxy analogues with which they are isoelectronic.

(iii) *Reaction of* $Fe_4C(CO)_{13}$ *with hydride*

$$\underset{\mathbf{4}}{[Fe_4C(CO)_{13}]} + LiEt_3BH \longrightarrow \underset{\mathbf{7}}{Li^+[Fe_4(CO)_{12}C \cdot CHO]^-.} \quad (6)$$

The addition of one equivalent of lithium triethylborohydride in diethyl ether to an ether solution of **4** results in the formation of the lithium salt of the *C*-formyl derivative, **7**, which is apparently stable only as its ion-paired lithium salt ($\nu_{CO(formyl)} = 1560\ cm^{-1}$). Attempts to isolate salts with other cations result in decomposition to $[HFe_4C(CO)_{12}]^-$, but the reactivity of **7** with alkylating agents to give a neutral vinylidene cluster (see §5*d*) supports the formulation of **7** as $Li^+[Fe_4(CO)_{12}C \cdot CHO]^-$.

(iv) *Reaction of* $Fe_4C(CO)_{12}^{2-}$ *with alkylating agents*

$$\underset{\mathbf{5}}{[Fe_4C(CO)_{12}]^{2-}} + RI \longrightarrow [Fe_4(CO)_{12}CC(O)R]^- + I^-. \quad (7)$$

Reaction of the dianion **5** with alkyl iodides in methylene chloride yields the *C*-acyl derivatives (reported previously for $R = CH_3$) (Davis *et al.* 1981), in a process that requires one mol of carbon monoxide. The reaction proceeds in only moderate yield, which suggests that the additional CO is provided by sacrificial reaction of some of the reagent molecules. No structural data are yet available.

(b) C–H *bond formation*

(i) *Reaction of* $Fe_4C(CO)_{13}$ *with dihydrogen*

$$\underset{\mathbf{4}}{Fe_4C(CO)_{13}} + H_2 \xrightarrow[\text{reflux}]{\text{toluene}} \underset{\mathbf{8}}{HFe_4(CO)_{12}CH} + CO. \quad (8)$$

The thermal decarbonylation of $Fe_4C(CO)_{13}$ in refluxing toluene generates coordinatively unsaturated $[Fe_4C(CO)_{12}]$, which then reacts with dihydrogen to yield the hydrido methylidyne cluster, **8**. This compound was previously synthesized by Tachikawa & Muetterties (1980) by the reaction of $[Fe_4C(CO)_{12}]^{2-}$ with silver ion under hydrogen (probably via the same intermediate) and was fully characterized by X-ray (Beno *et al.* 1980) and neutron diffraction (Beno *et al.* 1981). The outstanding structural feature of the butterfly shaped molecule is a three-centre FeHC interaction involving a wingtip iron atom and the methylidyne group. The formation of the C–H group by reaction between the cluster bound carbon atom and hydrogen is highly significant in that it provides a molecular analogue for the reaction of surface-bound carbon atoms with hydrogen in heterogeneously catalysed methanation and Fischer–Tropsch catalysis.

It is noteworthy that the methylidyne hydrogen in **8** is quite acidic, and ionization in methanol is quantitative, yielding $[HFe_4C(CO)_{12}]^-$, **6** (Holt *et al.* 1981). The protonation of the dianion **5** to **6** by one equivalent of acid reflects the low basicity of the exposed carbon atom in **5**.

The reactivity of the carbon atom raises questions about its chemical nature: should it be regarded as cationic, neutral (carbenoid perhaps) or anionic, as the name carbidocarbonyl has long implied? No mechanistic data are available for the reactions at the carbon atom, and so any observations we can make on the nature of the carbon in its cluster-bound environment must remain speculative. We have suggested (Bradley *et al.* 1979) that the reaction between alcohols (and amines) with $Fe_4C(CO)_{13}$ may proceed via a ketenylidene intermediate $Fe_4(CO)_{12}{:}C{:}CO$. Although there is no direct evidence for this, the reactions of $H_2Os_3(CO)_{10}CCO$ and $Co_3(CO)_9$-CCO^+ with alcohols to give $H_3Os_3(CO)_{12}C{\cdot}CO_2R$ (Sievert *et al.* 1982) and $Co_3(CO)_9CCO_2R$ (Seyferth *et al.* 1974) provide support for this route. These reactions may be interpreted in terms of a positively charged carbon atom, by analogy with the similar reactions of carbonium ions with carbon monoxide and alcohols to give esters. Consistent with this proposal, we have observed that increasing the electron density on the cluster by substituting trimethylphosphine for carbon monoxide (see above) prevents the formation of the carbomethoxymethylidyne cluster:

$$Fe_4C(CO)_{13} \xrightarrow{\text{MeOH}} [Fe_4(CO)_{12}C{\cdot}CO_2Me]^-$$
$$\downarrow \text{PMe}_3$$
$$Fe_4C(CO)_9(PMe_3)_4 \xrightarrow{\text{MeOH}} \text{no reaction.} \tag{9}$$

Few spectroscopic data are available that might shed light on the electronic nature of the lone carbon atom in these clusters. In an attempt to gain some insight into this property, we have measured the ^{13}C n.m.r. chemical shifts of the carbide carbon atoms in a series of iron carbidocarbonyl clusters (see table 2). In every case, the carbide resonances are shifted strongly to low field and represent the largest ^{13}C shifts yet reported for diamagnetic organometallic molecules. This might be interpreted in terms of a deshielded positively charged carbon atom, but this is by no means an indisputable correlation. There seems to be no connection between the charge on the molecule and $\delta_{C_{\text{carbide}}}$. In the series $[Fe_4C(CO)_{13}]$, $[HFe_4C(CO)_{12}]^-$ and $[Fe_4C(CO)_{12}]^{2-}$, which are very similar in geometry (see figure 5) the carbide resonances are found at 469.8, 464.2 and 478.0×10^{-6}. The cause of the shifts observed for cluster-bound carbon atoms is not clear and may reflect the predominance of paramagnetic contributions to the chemical shift rather than simply deshielding at a cationic centre. Temperature-dependent solid-state ^{13}C n.m.r. studies are in progress to resolve this question, and more illuminating evidence on the nature of the carbon atom in this unique environment must await further spectroscopic and theoretical investigations, which are so far lacking for these molecules.

TABLE 2. $\delta_{C_{carbide}}$ FOR CARBIDOCARBONYL CLUSTERS OF IRON

(Recorded at 22.5 MHz on a Jeol FX-900 spectrometer, at probe temperature. All samples were enriched to *ca.* 20 % ^{13}C by ^{13}CO exchange with $Fe(CO)_5$, which was used as the precursor for all the clusters synthesized.)

	$10^6\delta$
$Fe_4C(CO)_{13}$†	468.9
$HFe_4C(CO)_{12}^-$†	464.2
$Fe_4C(CO)_{12}^{2-}$†	478.0
$Fe_4C(CO)_9(PMe_3)_4$†	471.5
$Fe_5C(CO)_{15}$†	486.0
$Fe_6C(CO)_{16}^{2-}$‡	484.6

† CD_2Cl_2 solution; ‡ $(CD_3)_2CO$ solution.

5. THE CHEMISTRY OF CARBIDE-BASED ORGANOMETALLIC Fe_4 CLUSTERS

We have investigated the chemistry of some of the *C*-derivatized Fe_4C clusters described above, and these will be exemplified by the chemistry of the carbomethoxymethylidyne cluster.

(a) Reaction with hydrogen

$$[Fe_4(CO)_{12}C \cdot CO_2CH_3]^- \xrightarrow[2.8\,MPa,\ 120\,°C]{H_2,\ THF} CH_3CO_2CH_3 + [Fe_6C(CO)_{16}]^{2-}. \qquad (10)$$

$$\underset{3}{} \qquad\qquad\qquad\qquad\qquad\qquad\qquad \underset{2}{}$$

On treatment of **3** in THF with hydrogen at 2.8 MPa (400 lbf in^{-2}) and 120 °C, the four Fe–C bonds to the organic group are hydrogenolysed to give methyl acetate, and under the conditions of this reaction the iron species present agglomerate to **2** in a carbide-producing reaction. Because **2** was the starting material for the synthesis of **3**, the potential (as yet unrealized) exists for a catalytic synthesis of methyl acetate. Attempts to produce more than a stoichiometric amount of methyl acetate by treating **3** with carbon monoxide and hydrogen in methanol have proved unsuccessful, with the irreversible formation of **2**.

(b) Reaction with acid

$$[Fe_4(CO)_{12}CCO_2CH_3]^- + H^+ \xrightarrow[25\,°C]{hexane} [Fe_4C(CO)_{13}] + CH_3OH. \qquad (11)$$

$$\underset{3}{} \qquad\qquad\qquad\qquad\qquad\qquad\qquad \underset{4}{}$$

As described earlier, protonation of the *C*-carbomethoxy derivative cleaves the methoxy group as methanol and produces the parent carbidocarbonyl of the Fe_4 family $[Fe_4C(CO)_{13}]$ in high yield.

(c) Reaction with $BH_3 \cdot THF$

$$[Fe_4(CO)_{12}C \cdot CO_2CH_3]^- \xrightarrow[25\,°C]{BH_3 \cdot THF} [HFe_4C(CO)_{12}]^-. \qquad (12)$$

$$\underset{3}{} \qquad\qquad\qquad\qquad\qquad\qquad\qquad \underset{6}{}$$

The borane reduction of the carbomethoxy derivative **3** produces the monoanion **6** in high yield. This material was previously prepared by other routes (Davis *et al.* 1981; Holt *et al.* 1981).

(d) Reaction with alkylating agents

$$[Fe_4(CO)_{12}CCO_2CH_3]^- + (CH_3)_3O^+ \xrightarrow[25\ ^\circ C]{CH_2Cl_2} [Fe_4(CO)_{12}C{=}C(OCH_3)_2]. \qquad (13)$$

3   **9**

The carbonyl of the *C*-carbomethoxy group **3** is methylated by trimethyloxonium fluoborate to yield the neutral dimethoxyvinylidene cluster, **9**, which has been characterized crystallographically.

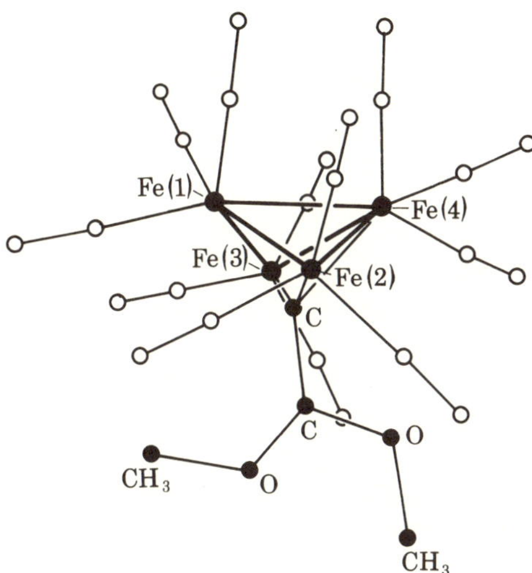

FIGURE 6. $Fe_4(CO)_{12}(C{=}C(OCH_3)_2)$, **9**. $Fe(1)$–$Fe(2)$ = $Fe(1)$–$Fe(3)$ = 2.491(3) Å; $Fe(2)$–$Fe(4)$ = $Fe(3)$–$Fe(4)$ = 2.555(2) Å; $Fe(1)$–$Fe(4)$ = 2.532(2) Å. $Fe(1)$–C = 1.976(9) Å; $Fe(2)$–C = 2.026(9) Å; $Fe(3)$–C = 2.028(9) Å; $Fe(4)$–C = 1.960(9) Å, vinylidene $C{=}C$ = 1.434(14) Å. Dihedral at $Fe(1)$–$Fe(4)$ = 127°, Angle $Fe(2)CFe(3)$ = 149°.

The structure of **9** is shown in figure 6. The open Fe_4 framework remains, with three terminal carbonyls on each metal atom. The geometry of the $[Fe_4(CO)_{12}C]$ core has altered little from that of the anionic precursor **3** (see figure 1). The dihedral angle at $Fe(1)$–$Fe(4)$ between the two triangular wings of the butterfly is 127° (cf. 130° in **3**) and the methylidyne carbon in **3** has not moved significantly in becoming the vinylidene carbon in **9** (angle $Fe(2)CFe(3)$ = 149°, cf. 148° for the corresponding angle in **3**). The dimethoxy vinylidene group is orientated in a plane that also includes the central iron–iron bond $Fe(1)$–$Fe(4)$ of the butterfly. Analogous reactions of trimethyloxonium fluoroborate with the *C*-acetyl and *C*-formyl clusters anions yield $Fe_4(CO)_{12}C{=}C(CH_3)OCH_3$ and $Fe_4(CO)_{12}C{=}C(H)OCH_3$, respectively.

6. SUMMARY

The chemistry of the Fe_4C family of clusters is summarized in scheme 1.

The synthesis and chemistry of these molecules is providing a rich and growing area of organometallic cluster chemistry. It remains to be seen whether or not the properties of this intriguing class of clusters will shed light on the reactivity of surface-bound carbon atoms on metallic heterogeneous catalysts. Such interdisciplinary analogies, while often stimulating and instructive, remain secondary to the fascination of the chemistry of the clusters themselves.

The extension of this area to the carbon atom chemistry in clusters of other metals than iron is a challenging goal, and its achievement, together with spectroscopic and theoretical investigations of the nature of the cluster-bound carbon atom, is eagerly awaited.

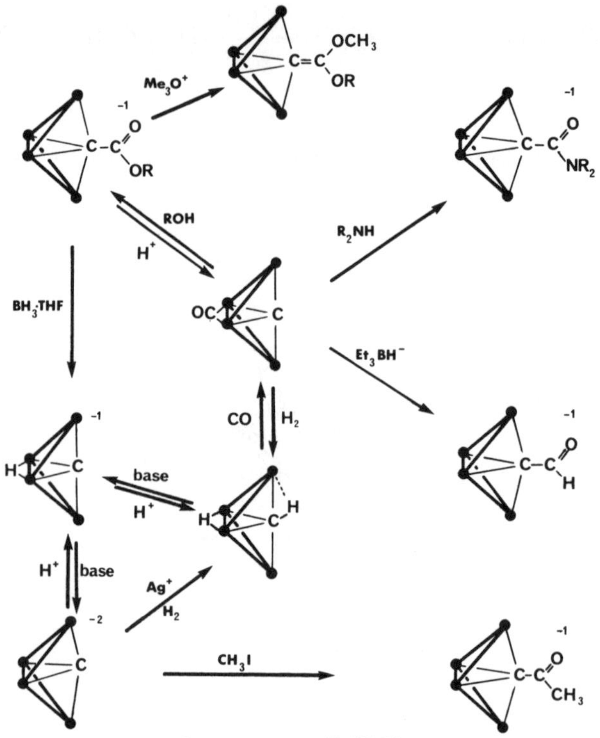

Scheme 1. ●, $Fe(CO)_3$.

The author acknowledges with gratitude the collaboration of E. W. Hill throughout the work described. Crystallographic results were provided by Dr G. B. Ansell, Dr M. E. Leonowicz and Ms M. Modrick of the Analytical and Information Division, Exxon Research and Engineering Company, and will appear in full under their name.

REFERENCES

Beno, M. A., Williams, J. M. & Muetterties, E. L. 1981 *J. Am. chem. Soc.* **103**, 1485–1492.

Beno, M. A., Williams, J. M., Tachikawa, M. & Muetterties, E. L. 1980 *J. Am. chem. Soc.* **102**, 4542–4544.

Bradley, J. S., Ansell, G. B. & Hill, E. W. 1979 *J. Am. chem. Soc.* **101**, 7417–7419.

Bradley, J. S., Ansell, G. B., Leonowicz, M. E. & Hill, E. W. 1981 *J. Am. chem. Soc.* **103**, 4968–4970.

Braye, E. H., Dahl, L. F., Hubel, W. & Wampler, D. L. 1962 *J. Am. chem. Soc.* **84**, 4633–4639.

Churchill, M. R., Wormald, J., Knight, J. & Mays, M. J. 1971 *J. Am. chem. Soc.* **93**, 3073–3074.

Davis, J. H., Beno, M. A., Williams, J. M., Zimmie, J., Tachikawa, M. & Muetterties, E. L. 1981 *Proc. natn. Acad. Sci. U.S.A.* **78**, 668–671.

Holt, E. M., Whitmire, K. H. & Shriver, D. F. 1981 *J. organometall. Chem.* **213**, 125–137.

Johnson, B. F. G., Johnston, R. D. & Lewis, J. 1967 *Chem. Commun.*, p. 1057.

Johnson, B. F. G., Johnston, R. D. & Lewis, J. 1968 *J. chem. Soc.* A, pp. 2865–2868.

Muetterties, E. L. 1977 *Science, Wash.* **196**, 839–848.

Ponec, V. 1978 *Catal. Rev. Sci. Engng* **18**, 151–171.

Seyferth, D., Hallgren, J. E. & Eschbach, C. S. 1974 *J. Am. chem. Soc.* **96**, 1730–1737.

Sievert, A. C., Strickland, D. S., Shapley, J. R., Steinmetz, G. R. & Geoffroy, G. L. 1982 *Organometallics* **1**, 214–215.

Tachikawa, M. & Muetterties, E. L. 1980 *J. Am. chem. Soc.* **102**, 4541–4542.

Tachikawa, M. & Muetterties, E. L. 1981 *Prog. inorg. Chem.* **28**, 203–238.

Ugo, R. 1975 *Catal. Rev. Sci. Engng* **11**, 225–297.

Wade, K. 1972 *Inorg. nucl. Chem. Lett.* **8**, 559–563.

Discussion

E. W. RANDALL (*Chemistry Department, Queen Mary College, London, U.K.*). If one is allowed to use the concept of bond order in clusters, what are the indications concerning the bond orders for the iron–carbido carbon interactions? The chemical shift for ^{13}C is dominated by the second-order paramagnetic effect, as Dr J. Mason would agree, for most carbons and almost certainly therefore for the carbido carbon.

J. S. BRADLEY. It is clearly not simple to define bond orders for the metal–carbon bonds in these clusters. Molecular orbital calculations are being done by my colleague Dr S. Harris to provide a clearer understanding of the nature of the interactions between the iron atoms and the carbide carbon. Our initial results indicate that the carbide carbon carries a fractional negative charge, which rules out any interpretation of the large downfield ^{13}C chemical shifts in terms of simple deshielding at a cationic centre.

Phil. Trans. R. Soc. Lond. A **308**, 115–124 (1982)   [115]
Printed in Great Britain

Catalysis by supported clusters: two examples of catalytic reactions that involve a molecular cluster frame

By J. M. Basset, B. Besson, A. Choplin and A. Theolier

Institut de Recherches sur la Catalyse, 2 avenue Albert Einstein,
69626 Villeurbanne Cédex, France

Two examples of catalysis by supported clusters are given that involve a molecular cluster frame during the catalytic cycle.

$(H)Os_3(CO)_{10}(OSi\leqslant)$ is able to hydrogenate ethylene between 343 and 373 K. The mechanism of this reaction, which involves the molecular cluster frame, requires the opening of the bridging three-electron oxygen ligand as the key step of the catalytic process.

$Rh_6(CO)_{16}$ adsorbed on alumina is a catalyst for the water-gas shift reaction carried out between 293 and 393 K. The mechanism of this reaction involves three steps: (i) electrophilic attack by surface protons on the metallic frame with formation of $Rh^I(CO)_2$ and $Rh^{III}(H)(H)$ species; (ii) reductive elimination of H_2 favoured by CO coordination to the $Rh^{III}(H)(H)$ species; (iii) nucleophilic attack by molecular water on CO coordinated to Rh^I with regeneration of the starting cluster, release of CO_2 and surface protons.

Introduction

Catalysis by molecular clusters (Smith & Basset 1977; Laine 1982; Jackson *et al.* 1982) remains of great interest in the field of homogeneous and heterogeneous catalysis. Among the various reasons for such an interest there are the 'cluster–surface analogy' (Johnson & Lewis 1975; Muetterties *et al.* 1979; Johnson 1981; Ugo 1975; Muetterties 1975a, b; Basset & Ugo 1977; Primet *et al.* 1975), the availability of an increasing number of new clusters of increasing nuclearity (Eady *et al.* 1975; Jackson *et al.* 1980; Chini 1980), and the increasing number of molecular cluster carbonyls used as precursor (or effective) (Chini 1980; Pruett & Walker 1974; Robertson & Webb 1974; Anderson & Howe 1977) catalysts in 'CO catalysis' or related reactions. Consider also the great variety of catalytic reactions that can be carried out on metal surfaces and that are not yet known for mononuclear complexes. It appears that one of the main motivations for trying to develop the field of cluster catalysis is to carry out selectively, at a molecular level, catalytic reactions that occur on metal surfaces but with poor selectivities.

Initially, the immobilization of a cluster onto a support, followed by a thermal or photochemical decarbonylation, appeared to be a way to transform a molecular cluster carbonyl 'poisoned' by its carbonyl ligands into a very small particle of metal (Robertson & Webb 1974; Smith *et al.* 1975, 1978; Ichikawa 1976; Anderson & Howe 1977; Kuznetzov *et al.* 1980; Knözinger & Rumpf 1978; Lieto & Gates 1980a, b). In fact such an approach proved to be much more complicated than expected (Smith *et al.* 1979; Bilhou *et al.* 1978; Besson *et al.* 1980; Psaro *et al.* 1981). The direct immobilization of a molecular cluster carbonyl onto the surface of an inorganic oxide has contributed to the recent development of two different fields, which can be called 'surface organometallic chemistry' and 'catalysis by supported clusters'.

In the former field one considers, *inter alia*, the reactions of the metallic frame or of the ligands with the functional groups present at the surface of an oxide such as the hydroxyl groups, the Lewis centres, and the siloxane or aloxane bridges. It is only after a careful study of the immobilized system or the reaction product of a cluster with a surface that it is possible to consider whether any catalytic reaction involves a cluster catalyst. Cluster catalysis on surfaces may then be compared with clusters in solution, i.e.: (i) the catalytic reaction involves a molecular metal cluster frame throughout the complete catalytic cycle; (ii) the catalytic reaction involves a molecular cluster in some steps of the catalytic cycle and mononuclear fragments in other steps; (iii) the catalytic reaction involves a metal particle derived from the molecular cluster. This last has been the most commonly encountered situation (Smith & Basset 1977).

We shall present two examples where the two first situations have been encountered, namely: (i) the catalytic hydrogenation of ethylene with $HOs_3(CO)_{10}(O\text{-}Si\leqslant)$, and (ii) the catalytic water-gas shift reaction carried out with $Rh_6(CO)_{16}$ adsorbed on alumina.

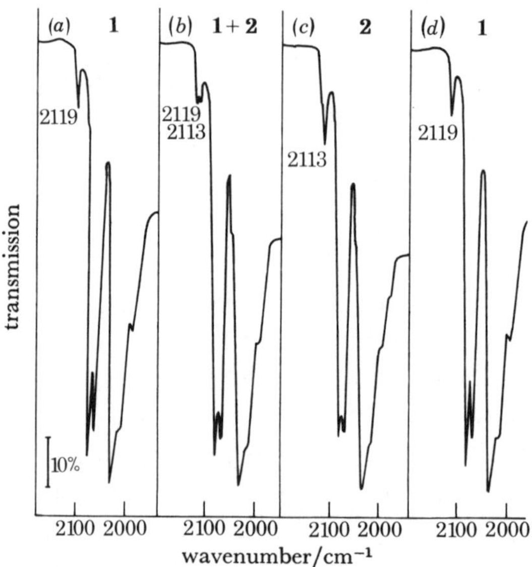

FIGURE 1. (a) Infrared spectrum in the ν(CO) region of $HOs_3(CO)_{10}(OSi\leqslant)$. (b) As (a), after treatment under C_2H_4 (33 kPa) at 333 K for 1 h. (c) As (b), after treatment at 353 K for 1 h. (d) As (c), after vacuum (10 mPa) after treatment at 353 K.

CATALYTIC HYDROGENATION OF ETHYLENE WITH $HOs_3(CO)_{10}(O\text{-}Si\leqslant)$

The reaction between $Os_3(CO)_{12}$ or $Os_3(CO)_{10}(CH_3CN)_2$ and the silanol groups of silica gives at 423 K the grafted cluster $(H)Os_3(CO)_{10}(OSi\leqslant)$ (1) (Besson *et al.* 1980; Psaro *et al.* 1981). This is a compound stable in air at room temperature and, under Ar, up to 473 K.

The hydrogenation of ethylene can be carried out with **1** between 343 and 373 K, in either batch or dynamic reactors. In a typical experiment, 100 mg of catalyst (1–SiO_2; 2% Os by mass) were introduced in a flow reactor. After a purge with He, the catalyst was heated at 363 K under a mixture of H_2 (50.5 kPa), C_2H_4 (7.2 kPa) and He (42 kPa) with a flow rate of 0.66 ml s⁻¹. A steady-state activity was obtained without any break-in period with a turnover of 40 mmol per mole of cluster per second (differential conditions). The orders with respect to the olefin and to the hydrogen were found to be respectively 0 and 1. The apparent activation energy, determined between 343 and 363 K, was found to be 32 kJ mol^{-1}. The i.r. spectrum of the catalyst after several runs was identical to that of **1**, suggesting that **1** was the effective catalyst. Mechanistic studies were therefore carried out on **1**.

Compound **1** is able to coordinate C_2H_4 at 353 K: volumetric measurements indicate that under 14.4 kPa C_2H_4, 20 % of **1** is transformed into an adduct with C_2H_4. This behaviour is confirmed by i.r. spectroscopy (figure 1*a–c*). Under 33 kPa C_2H_4, **1** is transformed almost quantitatively into a new compound, **2**, which gives back **1** under vacuum at 353 K (figure 1*d*).

The reversibility of the coordination process and the similarity of the i.r. spectra of **1** and **2** suggest that **2** has the structure $(H)(C_2H_4)Os_3(CO)_{10}(\mu_{-1}O\text{-}Si{\leqq})$ or $(C_2H_5)Os_3(CO)_{10}(\mu_{-2}O\text{-}Si{\leqq})$. Compound **1** does not coordinate H_2 at 353 K as determined by i.r. and volumetric measurements carried out at 273 or 353 K. However, under H_2 at 373 K, i.e. above reaction temperature, **1** gives compound **3** (figure 2*a, b*), which gives back **1** under vacuum at the same temperature (figure 2*c*). Compound **3** can also be easily obtained from **2** by H_2 treatment at 323 K (figure 3).

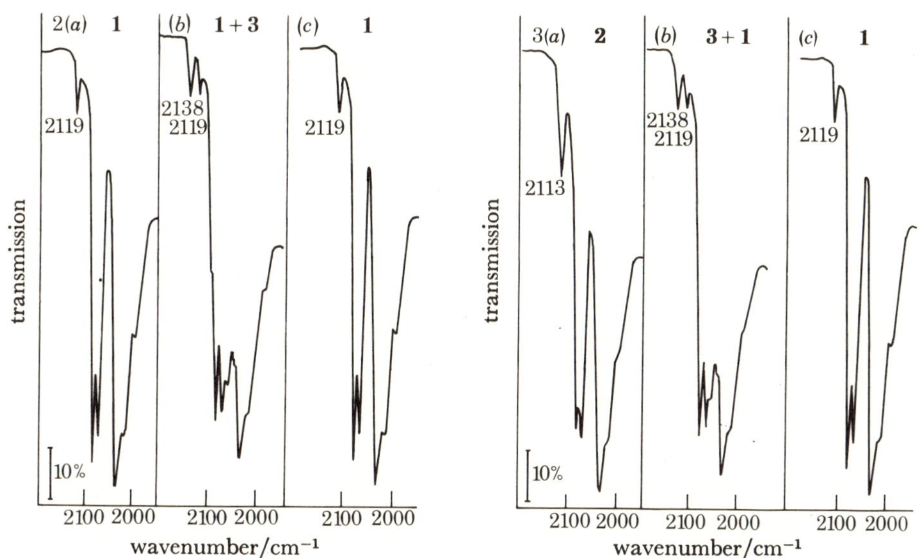

Figure 2. (*a*) Infrared spectrum in the $\nu(CO)$ region of **1**. (*b*) As (*a*), after H_2 treatment (32 kPa) at 373 K. (*c*) As (*b*), after vacuum treatment (10 mPa) at 273 K.

Figure 3. (*a*) Infrared spectrum in the $\nu(CO)$ region of **2**. (*b*) As (*a*), after H_2 treatment (32 kPa) at room temperature. (*c*) As (*b*), after vacuum treatment (10 mPa) at 273 K.

These results are interpreted in scheme 1. The coordination of the olefin at 353 K is likely to involve the opening of the Os–O–Os bridge, which creates a vacant coordination site on one metal atom. This process is similar to that reported with the bridging sulphur ligand of $Os_3(CO)_9H(SMe)$ (Johnson *et al.* 1978*a*), which coordinates C_2H_4 to give $Os_3(CO)_9H(C_2H_4)$ (SMe) (Johnson *et al.* 1978*b*). It is difficult at present to decide if the structure of the adduct compound contains the π-coordinated olefin (**2**) or the σ-alkyl group (**2'**). The reversibility of the olefin addition suggests only that if the σ-alkyl group is present, the β-H elimination is a facile and reversible process. The insertion process might be favoured by the concerted migration of the one-electron ligands σ-O and σ-H from the Os_1 to the Os_2 after π coordination of the olefin (scheme 1). The next step of the catalytic cycle involves the oxidative addition of H_2 on a coordinatively unsaturated Os_1 atom. The order of unity with respect to H_2 indicates that this step is rate-determining. The formation of **3** (hypothetical structure) from **1**, or from **2** or **2'**, indicates that H_2 can, under certain conditions, oxidatively add to the unsaturated cluster **1'**, which can be obtained either from **2** or **2'** after oxidative addition of H_2 and reductive elimination of C_2H_6 or from **1** by thermal activation under H_2. In the absence of C_2H_4 or after total consumption of C_2H_4, **1'** gives **3** under H_2.

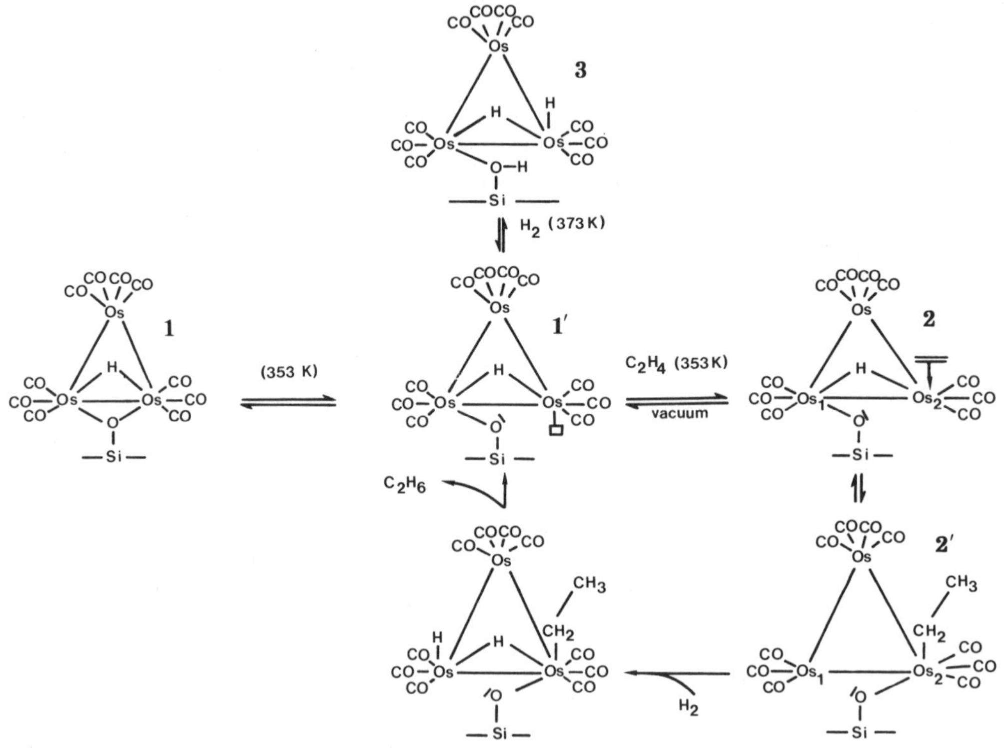

SCHEME 1

This example constitutes, to our knowledge, the first case of catalysis by a molecular cluster grafted onto unmodified silica. With functionalized silica, there is one reported case of ethylene hydrogenation and but-1-ene isomerization with the clusters $Os_3(CO)_{11}(PPh_2\text{-}SiL)$ and $H_2Os_3(CO)_9(PPh_2\text{-}SiL)$ (Brown & Evans 1981; Freeman *et al.* 1982). In our case the presence of a bridging hydride and of a bridging oxygen would be responsible for the appearance of catalytic properties. The absence of ageing of such a catalyst is probably due to grafting, which may stabilize coordinative unsaturation and avoid secondary reactions.

THE WATER-GAS SHIFT REACTION CATALYSED BY $Rh_6(CO)_{16}$ ADSORBED ON η-ALUMINA

A great variety of molecular cluster carbonyls have been used as precursor catalysts or as catalysts in the water-gas shift reaction (Ungermann *et al.* 1979; Ford *et al.* 1978; Laine *et al.* 1977; Kang *et al.* 1977; Marnot *et al.* 1981) or related reactions (Laine *et al.* 1978; Laine 1978;

Niwa *et al.* 1979; Cole *et al.* 1980).[†] In some cases catalytic cycles were proposed on spectroscopic evidence that suggested that a cluster frame could be retained during the catalytic reaction (Ford 1981). We report here that $Rh_6(CO)_{16}$, when supported on alumina, is an effective catalyst for the water-gas shift reaction; the mechanism of this reaction involves the destruction of the cluster frame and its reconstruction as two of the major steps of the catalytic cycle.

In a typical experiment 2 g η-alumina (315 m^2 g^{-1}) was degassed under vacuum (10 mPa) at 773 K for 15 h $((Al_2O_3)_{773})$. The resulting solid was contacted in a vacuum line with a solution of $Rh_6(CO)_{16}$ (30 µmol) in CH_2Cl_2. After removal of the solvent under vacuum, the resulting catalyst (**4**) was contacted with 2.9 kPa H_2O and 39 kPa CO in a glass reactor. The water-gas shift reaction was observed by the appearance of equivalent amounts[‡] of gaseous CO_2 and H_2 at temperatures as low as 298 K and up to about 373 K, temperature at which the lifetime of the catalyst was very short (less than 2 h). With **4**, for a temperature of 353 K, the initial turnover number was equal to 2 mol H_2 or CO_2 per mole of $Rh_6(CO)_{16}$ per hour (figure 4). At 323 K the order with respect to CO and H_2O were respectively 1 (figure 5) and 0. The apparent activation energy, taken in the temperature range 298–353 K and for a CO pressure of 39 kPa was 65 kJ mol^{-1}. The catalyst aged with time at temperatures equal or above *ca.* 323 K or at CO pressures below *ca.* 20 kPa.

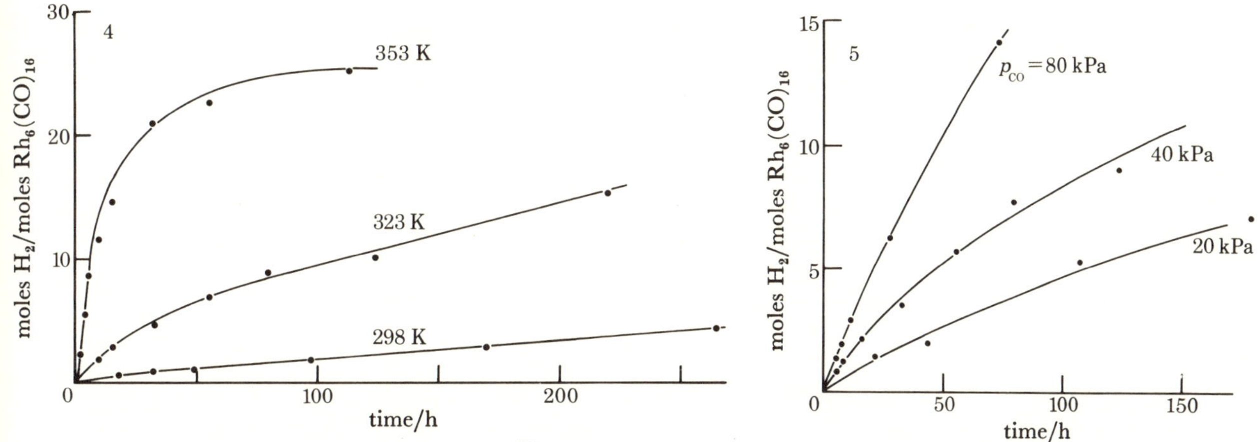

FIGURE 4. Typical kinetic results of the water-gas shift reaction obtained with catalyst **4** at various temperatures (batch reactor) (p_{CO} = 40 kPa).

FIGURE 5. Influence of the initial CO pressure on the initial rate of the water-gas shift reaction (catalyst **6**; T = 323 K).

The reactivity of the molecular cluster adsorbed onto η-alumina in the presence of CO or H_2O, or both, was studied by infrared spectroscopy and analysis of the gas phase. When $Rh_6(CO)_{16}$ was adsorbed on $(Al_2O_3)_{773}$, the resulting infrared spectrum, after removal of the CH_2Cl_2 solvent, exhibited $\nu(CO)$ bands at 2086 (vs) and 1845 (br, m) cm^{-1}. The cluster was only slightly interacting with the partly dehydroxylated surface, because it could be removed

---

† There are also a number of mononuclear complexes that catalyse the water-gas shift reaction under alkaline or acidic conditions (see, for example, Cheng *et al.* 1977; Cheng & Eisenberg 1978; Yoshida *et al.* 1978; King *et al.* 1980; Baker *et al.* 1980).

‡ The amounts of CO_2 and H_2 were nearly equivalent within 5 % of experimental error. However, in the early stages of the reaction the amount of CO_2 was slightly lower than that of H_2, presumably due to the absorption of CO_2 on the hydrated support (carbonate).

from the surface as $Rh_6(CO)_{16}$ by washing with CH_2Cl_2. If 20 mg of the alumina-supported cluster was treated with 5 µmol H_2O for 5 min, so that the coverage of the alumina by OH groups was close to but less than unity, the cluster was transformed instantaneously into two new carbonyl compounds, **5** and **6**, characterized respectively by a doublet of equal intensity at 2094 and 2026 cm^{-1}, and a single band of smaller intensity at 2115 cm^{-1}. Simultaneously about 0.5 mol of H_2 per mole of cluster and about 5 mol of CO per mole of cluster were evolved without any gaseous CO_2 or bands that could arise from the adsorption of CO_2 on the alumina support, when carbonate species are observed by bands at 1222, 1450, 1480 and 1640 cm^{-1}. The $\nu(CO)$ band corresponding to the carbonyl compound **6** was eliminated under vacuum at 353 K. It is probably due to CO coordinated to a Rh^{III} surface complex. Treatment of the resulting solid with 39 kPa CO resulted in a slow and significant increase of the intensity of the doublet at 2094 and 2026 cm^{-1}. Simultaneously H_2 was slowly evolved with time (figure 6).

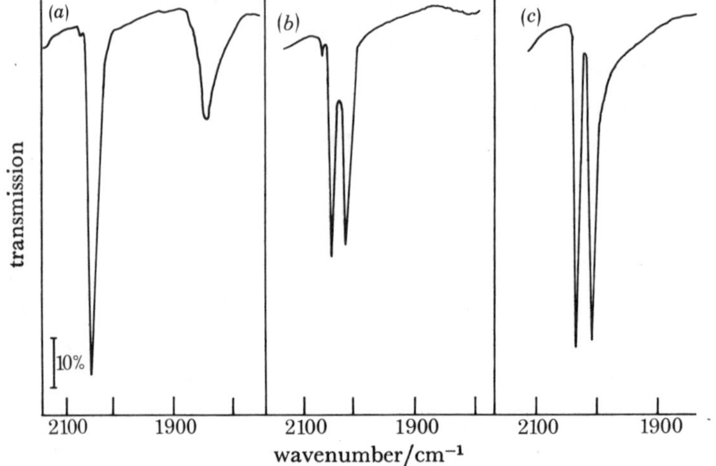

FIGURE 6. (a) Infrared spectrum in the $\nu(CO)$ region of $Rh_6(CO)_{16}$ adsorbed on $(Al_2O_3)_{773}$ (20 mg). (b) As (a), after treatment of the catalyst with 5 µmol H_2O for 5 min. (c) As (b), after treatment of the catalyst with 39 kPa CO for 24 h.

Compound **5** could also be obtained by reacting $[Rh(CO)_2Cl]_2$ with the surface of an alumina previously degassed at 573 K either from CH_2Cl_2 or from pentane solution, which suggests that this species is of the type $(CO)_2Rh^I(O–Al <)$ (Smith *et al.* 1979). Such carbonyl species (**2**) generated either from $Rh_6(CO)_{16}$ adsorbed on hydroxylated alumina or from $[Rh(CO)_2Cl]_2$ can be reversibly decarbonylated by treatment under vacuum (10 mPa) at 673 K and then with CO (40 kPa) at 273 K. Compound **5** does not oxidatively add H_2 at room temperature as determined by volumetric measurement and infrared spectroscopy.

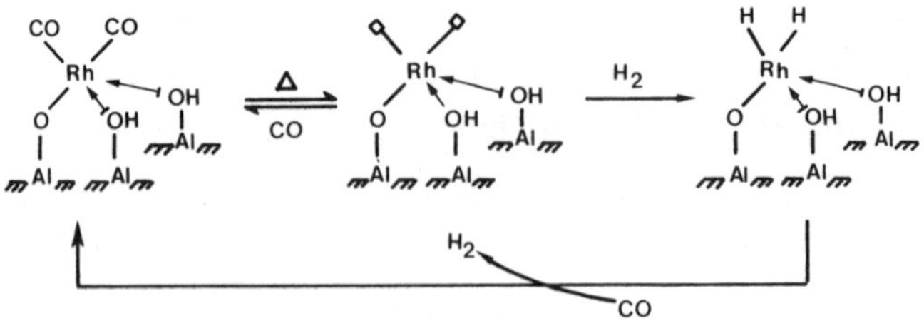

SCHEME 2

However, **5**, after decarbonylation at 673 K, is able to coordinate hydrogen reversibly at 298 K (0.62 mol H_2 per mole of Rh), under an H_2 equilibrium pressure of 39 kPa, which suggests the occurrence of a reversible process as depicted in scheme 2.

Compound **5** can also coordinate molecular water as determined by the small but significant shift of frequency observed by the adsorption of 2.9 kPa H_2O at 298 K: $\nu(CO)$ bands are shifted respectively from 2094 and 2026 cm^{-1} to 2089 and 2019 cm^{-1}. This process probably involves the substitution of the π donating –OH or –O– ligands of the surface coordinated to the rhodium I dicarbonyl by molecular water.

If compound **5** derived from $Rh_6(CO)_{16}$ adsorbed on hydroxylated alumina or from $[Rh(CO)_2Cl]_2$ was treated with CO (100 kPa) and H_2O (2.9 kPa), the resulting spectrum changed slowly with time to give a new carbonyl complex with $\nu(CO)$ bands at 2070 and 1804 cm^{-1}, typical of $Rh_6(CO)_{16}$ adsorbed on a fully hydrated alumina (Smith *et al.* 1979) (figure 7). Other bands at 1222, 1450, 1480 and 1640 cm^{-1}, corresponding to CO_2 adsorbed on alumina, were observed along with gaseous CO_2. The molecular cluster could then be extracted from the surface by $CHCl_3$ and characterized by its infrared spectrum. If the CO pressure was kept below *ca.* 20 kPa, the transformation of **5** into $Rh_6(CO)_{16}$ was incomplete and $\nu(CO)$ bands characteristic of CO coordinated to metallic rhodium (Smith *et al.* 1979) were observed at *ca.* 2025 (s) and 1850 (br.) cm^{-1}. Under catalytic conditions at temperatures between 298 and 353 K, and for CO pressures between 20 and 90 kPa, the infrared spectrum of the system indicated the presence of a $(CO)_2Rh^I(OAl{<})$ species and $Rh_6(CO)_{16}$.

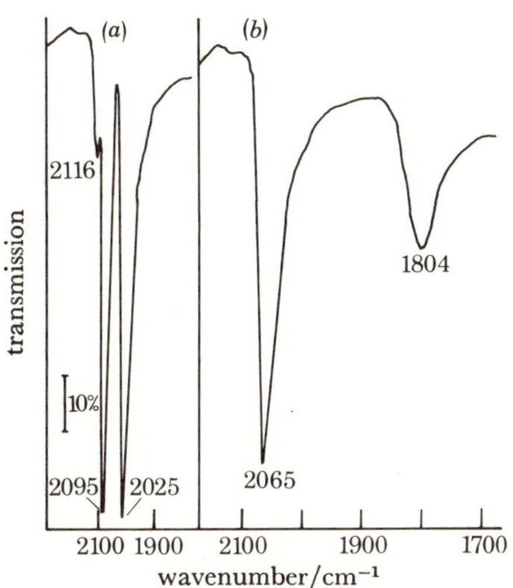

FIGURE 7. (a) Infrared spectrum in the $\nu(CO)$ region of the catalyst resulting from the treatment by a small amount of water of $Rh_6(CO)_{16}$ adsorbed on $(Al_2O_3)_{773}$. (b) As (a), after treatment with CO (100 kPa) and H_2O (2.9 kPa) for 24 h.

It was not possible to detect by infrared spectroscopy any rhodium hydride species; this may be due to the low intensity expected for the Rh–H or Rh–D vibrations. However, the $\nu(CO)$ frequency of **6** corresponded to CO coordinated to a Rh^{III} species that could be formulated as $Rh^{III}(CO)(H)(H)(O-Al{<})$.[†] The presence of a rhodium hydride was also demonstrated by the following experiment: 25 mmol $Rh_6(CO)_{16}$ supported on $(Al_2O_3)_{773}$ were treated by

---

[†] A similar species has been obtained on silica by Ward & Schwartz (1981) by reacting $Rh(\pi \text{ allyl})_3$ with silanol groups of silica. In his case the $(\nu Rh–H)$ band could be observed more easily than in ours, owing to the absence of $\nu(CO)$ bands.

0.4 mmol D_2O. This solid, when treated for 1 h at 373 K with excess C_2H_4, gave 30 µmol of $C_2H_4D_2$ and smaller amounts of C_2H_5D and C_2H_6; 50 µmol of H_2 would have been expected from the complete reaction depicted in scheme 3.

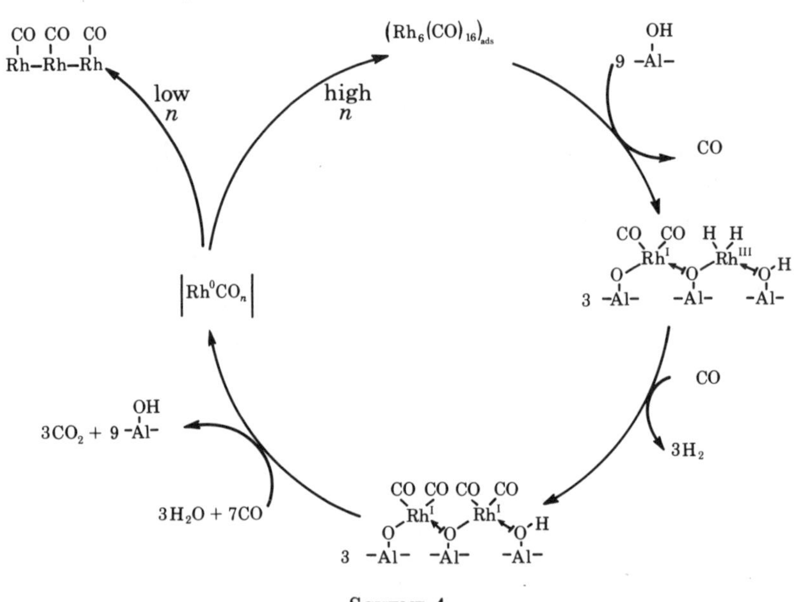

SCHEME 3

SCHEME 4

Scheme 4 rationalizes the catalytic cycle that occurs in the temperature range 293–393 K. The primary step would be the oxidative addition of Al–OH groups of the alumina to the cluster frame with the formation of $Rh^I(CO)_2$ and $Rh^{III}(H)(H)$ species. CO could coordinate to the $Rh^{III}(H)(H)$ species. The slow step of the process seems to be the expulsion of dihydrogen from the $Rh^{III}(H)(H)$ species. It is likely that this process involves the replacement by CO of water molecules or of OH groups or aloxane groups π coordinated to the $Rh^{III}(H)(H)$ species. The next step of the process seems to be a nucleophilic attack by molecular water on CO coordinated to the $Rh^I(CO)_2$ species with regeneration of a $Rh^0(CO)_n$ species and of a proton, as an Al–OH group.

The $Rh^0(CO)_n$ species may then regenerate the starting cluster. This occurs at low temperature ($T < 393$ K) and at high $CO:H_2O$ ratios. Otherwise the $Rh^0(CO)_n$ species agglomerates to metal particles, which are inactive in the catalytic reaction. This process leading to the formation of metal particles seems to be the origin of the ageing of the catalyst.

References

Anderson, J. R. & Howe, R. F. 1977 *Nature, Lond.* **268**, 129–130.

Baker, E. C., Hendriksen, D. E. & Eisenberg, R. 1980 *J. Am. chem. Soc.* **102**, 1020–1027.

Basset, J. M. & Ugo, R. 1977 In *Aspects of homogeneous catalysis* (ed. R. Ugo), vol. 3, pp. 137–183. Dordrecht: D. Reidel.

Besson, B., Moraweck, B., Smith, A. K., Basset, J. M., Psaro, R., Fusi, A. & Ugo, R. 1980 *J. chem. Soc. chem. Commun.*, pp. 569–571.

Bilhou, J. L., Bougnol, V., Graydon, W. F., Smith, A. K., Zanderighi, G. M., Basset, J. M. & Ugo, R. 1978 *J. organomettal. Chem.* **153**, 73–84.

Brown, S. C. & Evans, J. 1981 *J. molec. Catal.* **11**, 143–149.

Cheng, C. H. & Eisenberg, R. 1978 *J. Am. chem. Soc.* **100**, 5968–5970.

Cheng, C. H., Hendriksen, D. E. & Eisenberg, R. 1977 *J. Am. chem. Soc.* **99**, 2791–2792.

Chini, P. 1980 *J. organometall. Chem.* **200**, 37–61.

Cole, T., Ramage, R., Cann, K. & Pettit, R. 1980 *J. Am. chem. Soc.* **102**, 6182–6184.

Eady, C. R., Johnson, B. F. G. & Lewis, J. 1975 *J. chem. Soc. Dalton Trans.*, pp. 2606–2611.

Ford, P. 1981 *Acct. chem. Res.* **14**, 31–37.

Ford, P., Ringer, R. G., Ungermann, C., Laine, R., Landis, V. & Noya, S. A. 1978 *J. Am. chem. Soc.* **100**, 4595–4597.

Freeman, M. B., Patrick, M. A. & Gates, B. C. 1982 *J. Catal.* **73**, 82–90.

Ichikawa, M. 1976 *J. chem. Soc. chem. Commun.*, pp. 26–27.

Jackson, P. F., Johnson, B. F. G. & Lewis, J. 1980 *J. chem. Soc. chem. Commun.*, pp. 224–226.

Jackson, S. D., Wells, P. B., Whyman, R. & Worthington, P. 1982 In *Catalysis (Spec. Per. Report)*, vol. 4, pp. 75–99. London: Royal Society of Chemistry.

Johnson, B. F. G. 1981 *Platinum Metals Rev.* **25**, 62–70.

Johnson, B. F. G. & Lewis, J. 1975 *Pure appl. Chem.* **44**, 43–82.

Johnson, B. F. G., Lewis, J. & Pippard, D. 1978 *J. organometall. Chem.* **160**, 263–274.

Johnson, B. F. G., Lewis, J., Pippard, D. & Raithby, P. R. 1978*b* *J. chem. Soc. chem. Commun.*, pp. 551–552.

Kang, H. C., Mauldin, C. H., Cole, T., Slegeir, W., Cann, K. & Pettit, R. 1977 *J. Am. chem. Soc.* **99**, 8323–8325.

King, A. D., King, R. B. & Yang, D. B. 1980 *J. chem. Soc. chem. Commun.*, pp. 529–530.

Knözinger, H. & Rumpf, E. 1978 *Inorg. chim. Acta* **30**, 51–58.

Kuznetzov, V., Bell, A. T. & Yermakov, Y. I. 1980 *J. Catal.* **65**, 374–389.

Laine, R. 1978 *J. Am. chem. Soc.* **100**, 6451–6454.

Laine, R. M. 1982 *J. molec. Catal.* **14**, 137–169.

Laine, R. M., Ringer, R. G. & Ford, P. 1977 *J. Am. chem. Soc.* **99**, 252–253.

Laine, R. M., Thomas, D. W., Cary, L. W. & Buttrill, S. E. 1978 *J. Am. chem. Soc.* **100**, 6527–6528.

Lieto, J. & Gates, B. C. 1980*a* *Chemtech*, pp. 195–199.

Lieto, J. & Gates, B. C. 1980*b* *Chemtech*, pp. 248–251.

Marnot, P. A., Ruppert, R. R. & Sauvage, J. P. 1981 *Nouv. J. Chim.* **5**, 543–545.

Muetterties, E. L. 1975*a* *Bull. Soc. chim. Belg.* **84**, 959–986.

Muetterties, E. L. 1975*b* *Bull. Soc. chim. Belg.* **85**, 451–470.

Muetterties, E. L., Rhodin, T. N., Band, E., Brucker, C. G. & Pretzer, W. R. 1979 *Chem. Rev.* **79**, 91–137.

Niwa, M., Iizuka, T. & Lunsford, J. 1979 *J. chem. Soc. chem. Commun.*, pp. 684–685.

Primet, M., Basset, J. M., Garbowski, E. & Mathieu, M. V. 1975 *J. Am. chem. Soc.* **97**, 3655–3659.

Pruett, R. L. & Walker, W. E. 1974 U.S. Patents nos 3833 634 and 3937 857.

Psaro, R., Ugo, R., Zanderighi, G. M., Besson, B., Smith, A. K. & Basset, J. M. 1981 *J. organometall. Chem.* **213**, 215–247.

Robertson, J. & Webb, G. 1974 *Proc. R. Soc. Lond.* A **341**, 383–398.

Smith, A. K. & Basset, J. M. 1977 *J. molec. Catal.* **2**, 229–241.

Smith, A. K., Hugues, F., Theolier, A., Basset, J. M., Ugo, R., Zanderighi, G. M., Bilhou, J. L., Bougnol, V. & Graydon, W. F. 1979 *Inorg. Chem.* **18**, 3104–3112.

Smith, A. K., Theolier, A., Basset, J. M., Commereuc, D., Chauvin, Y. & Ugo, R. 1978 *J. Am. chem. Soc.* **100**, 2590–2591.

Smith, G. C., Chosnacki, T. P., Dasgupta, S. R., Iwatate, K. I. & Watters, K. L. 1975 *Inorg. Chem.* **14**, 1419–1421.

Ugo, R. 1975 *Catal. Rev.* **2**, 225–297.

Ungermann, C., Landis, V., Moya, S. A., Cohen, H., Walker, H., Pearson, R. G., Ringer, R. G. & Ford, P. 1979 *J. Am. chem. Soc.* **101**, 5922–5929.

Ward, M. D. & Schwartz, J. 1981 *J. molec. Catal.* **11**, 397–402.

Yoshida, T., Ueda, Y. & Otsuka, S. 1978 *J. Am. chem. Soc.* **100**, 3941–3942.

Discussion

R. WHYMAN (*I.C.I. New Science Group, Runcorn, U.K.*). The three $\nu(CO)$ absorptions (two bands and one shoulder) assigned to the presence of mononuclear $Rh(CO)_2$ fragments on the surface of alumina might more preferably be accounted for in terms of dinuclear rhodium dicarbonyl units with a structure analogous to that of $[Rh(CO)_2Cl]_2$ (with chloride ions replaced by surface oxide ions). We obtained spectroscopic and chemical evidence consistent with such an interpretation with silica as support material. The occurrence on the surface of a $Rh_6 \rightleftharpoons 3Rh_2$ reversible rearrangement might perhaps be more plausible than a $Rh_6 \rightleftharpoons 6Rh_1$ transformation and of course it has a clear precedent in the solution chemistry of rhodium carbonyls, i.e.

$$\xrightarrow{\text{ increasing pressure}}$$

$$Rh_6(CO)_{16} \underset{}{\overset{CO}{\rightleftharpoons}} Rh_4(CO)_{12} \overset{CO}{\rightleftharpoons} Rh_2(CO)_8.$$

$$\xleftarrow{\text{ increasing temperature}}$$

Phil. Trans. R. Soc. Lond. A **308**, 125–130 (1982)   [125]
Printed in Great Britain

The homogeneous hydrogenation of carbon–carbon multiple bonds with heteronuclear platinum clusters

By A. Fusi[1], R. Ugo[1], R. Psaro[1], P. Braunstein[2] and J. Dehand[2]

[1] *Istituto di Chimica Generale e Inorganica, C.N.R. Centre,*
Via Venezian, 21, 20133 Milano, Italy
[2] *Institut de Chimie, Université Louis Pasteur, 1 Rue Blaise Pascal,*
76008 Strasbourg Cedex, France

The heteronuclear platinum clusters

$$Pt_2Co_2(CO)_8(PPh_3)_2 \quad \text{and} \quad Pt(C_6H_{11}NC)_2[Mo(CO)_3(\eta^5\text{-}C_5H_5)]_2$$

are poor catalysts for the hydrogenation of olefins and internal acetylenes, but they show a reasonable activity for the hydrogenation of terminal acetylenes. In this latter case the selectivity towards olefins is low.

Whereas the $PtMo_2$ cluster can be recovered unaltered from the reaction mixture, the Pt_2Co_2 cluster undergoes rather complex molecular rearrangements, which take place under the catalytic conditions.

Finally the Pt_2Co_2 cluster, under well defined conditions, may act as a catalyst for the selective hydrogenation of diphenylacetylene to *cis*-stilbene.

1. Introduction

In the last few years molecular metallic clusters have been proposed as models of very small metallic particles such as those of highly dispersed supported metallic catalysts (Basset & Ugo 1977). However, even for a relatively simple catalytic reaction such as the hydrogenation of C–C multiple bonds it was reported that the activity of polynuclear metal clusters is usually less than that found for mononuclear metal complexes or for heterogeneous metallic catalysts. In addition, firm evidence that the original metal frame of the cluster is maintained unaltered under the catalytic conditions is difficult to acquire (Laine 1982; Smith & Basset 1977).

Metal clusters act as homogeneous catalysts usually under rather severe conditions (e.g. high temperature, high pressure, or both); this low intrinsic activity can be attributed to various factors. A major cause of the lack of catalytic activity of metal clusters under mild conditions could be the coordinative saturation of the metal atoms in the cluster frame (Norton 1977; Laine 1982). With this in mind, we have carried out an investigation on the hydrogenation of unsaturated C–C bonds catalysed by some heteronuclear platinum clusters that have one or more formally coordinatively unsaturated platinum atoms. Examples are the butterfly structure of $Pt_2Co_2(CO)_8(PPh_3)_2$ and the linear structure of *trans*-PtL_2M_2, ($L = C_6H_{11}NC$, pyridine or pyridine derivatives, $M = Co(CO)_4$, $Fe(NO)(CO)_3$, $Mo(CO)_3(\eta^5\text{-}C_5H_5)$) or the triangular structure of $PtCo_2(CO)_7DPE$ (Braunstein & Dehand 1970; Braunstein *et al.* 1970, 1975; Dehand & Nennig 1974; Barbier & Braunstein 1978; Pearson & Dehand 1969).

The choice of mixed heteronuclear platinum clusters was prompted by the report that the formation of heteronuclear platinum–metal bonds (e.g. Pt–Sn bonds) enhances the properties of platinum complexes as hydrogenation catalysts (Coffey 1970).

2. Hydrogenation of olefins

We studied the clusters listed above under the following range of the reaction parameters: variation of temperatures from 25 to 100 °C; variation of hydrogen pressures up to 100 atm; variation of molar ratio of substrate to catalyst from 10 to 100 and variation of reaction times from few hours up to 40 h.

Table 1. Hydrogenations catalysed by $Pt_2Co_2(CO)_8(PPh_3)_2$

([Cat] = 3.3 ± 0.5 mM; [Sub]/[Cat] = 110 ± 10; $T = 50$ °C;
the solvent was toluene unless otherwise stated.)

substrate	reaction time/h	pressure/atm	percentage conversion[†]	percentage isomerization	percentage hydrogenation
1-octene	40	33	47	43	4
cyclohexene	40	35	6 ± 1	—	6 ± 1
cyclooctene[‡]	20	50	4 ± 1	—	4 ± 1
1,3-cyclooctadiene[‡]	20	50	1.5	—	1.3 (0.2)[§]
1,5-cyclooctadiene[‡]	20	50	6.4	2 (0.9)[¶]	3.3 (0.2)[§]
styrene	20	50	3–4	—	3–4[†]

† In some catalytic runs higher conversions (*ca.* 10–30 %) have been obtained, but data were not reproducible and a metallic solid separated.
‡ With benzene as solvent.
§ Double hydrogenation to alkane in parentheses.
¶ Isomerization to 1,4-cyclooctadiene in parentheses.

Table 2. Hydrogenations catalysed by $Pt(C_6H_{11}NC)_2[Mo(CO)_3(\eta^5\text{-}C_5H_5)]_2$

([Cat] = 3.0 ± 0.4 mM; $T = 50$ °C; reaction time = 20 h; [Sub] = 0.3 ± 0.04 M;
the solvent was toluene unless otherwise stated; pressure = 48 ± 2 atm.)

substrate	percentage conversion	percentage isomerization	percentage hydrogenation
styrene[†]	2.6	—	2.6
1,3-cyclooctadiene[‡]	2.9	—	2.9
1-octene[§]	8	6.5	1.5

† With a reaction time of 65 h the conversion rises to 10.5 %; under a pressure of 70 atm for 20 h the conversion rises to 13 ± 1 %.
‡ With benzene as solvent.
§ [Sub]/[Cat] = 50.

Some compounds (e.g. $PtL_2[Fe(NO)(CO)_3]_2$ and $PtL_2[Co(CO)_4]_2$, where L = $C_6H_{11}NC$ or pyridine, and $PtCo_2(CO)_7DPE$) do not show any catalytic activity for isomerization or hydrogenation of olefins under all the reaction conditions investigated. $Pt_2Co_2(CO)_8(PPh_3)_2$ (1) isomerizes 1-octene under 1 atm of hydrogen at room temperature. More drastic conditions increase both the hydrogenation and isomerization properties of 1 and $Pt(C_6H_{11}NC)_2[Mo(CO)_3 (\eta^5\text{-}C_5H_5)]_2$ (2). We have examined in detail the behaviour of these two catalysts by carrying out an extended investigation on the hydrogenation of non-activated and activated terminal monoolefins, of cyclic internal olefins and of cyclic diolefins under the following standard conditions: temperature about 50 °C, hydrogen pressure about 50 atm, molar ratio of substrate to catalyst about 100, reaction time between 20 and 40 h. The results are reported in tables 1 and 2. Both clusters are very poor hydrogenation catalysts towards all olefins studied, although they catalyse the isomerization of 1-octene to a major extent.

Catalyst **2**, with a $PtMo_2$ metallic frame, can be completely recovered; infrared and thin-layer investigations did not show formation of any other metal complexes. On the other hand, an extended transformation and rearrangement was found for catalyst **1**, which has the Pt_2Co_2 metallic frame. The examination of the nature of the metal complexes recovered after the catalytic runs showed the quantitative transformation of the original cluster **1** into the known platinum cluster $Pt_5(CO)_6(PPh_3)_4$ (Barbier *et al.* 1978) and a series of cobalt complexes.

We could not isolate other platinum complexes, even in trace quantities. This unusual kind of disproportionation of the original heteronuclear Pt_2Co_2 cluster cage is not only independent of the nature of the olefin, but occurs only in the presence of both excess of olefin and of hydrogen pressure. In fact cluster **1** remains rather stable when maintained under 50 atm H_2 at 50 °C for 20 h (but in the absence of olefin) or when the olefin is added (but the pressure of H_2 is replaced by a similar pressure of N_2).

TABLE 3. HYDROGENATIONS OF TERMINAL ACETYLENES

([Cat] $= 2.9 \pm 0.3$ mm; [Sub]/[Cat] $= 110 \pm 10$; $T = 50$ °C; pressure $= 50$ atm; time $= 20$ h; the solvent was toluene.)

catalyst	substrate	percentage conversion	percentage monohydrogenation	percentage dihydrogenation	percentage oligomerization
$Pt(C_6H_{11}NC)_2[Mo(CO)_3(\eta^5\text{-}C_5H_5)]_2$	$PhC{\equiv}CH$	98	63	33	2
$Pt(3Mepy)_2[Mo(CO)_3(\eta^5\text{-}C_5H_5)]_2$	$PhC{\equiv}CH$	45†	9.5	0.5	35
$Pt(C_6H_{11}NC)_2[Mo(CO)_3(\eta^5\text{-}C_5H_5)]_2$	$nC_6H_{13}C{\equiv}CH$	95	34	61	—
$Pt_2Co_2(CO)_8(PPh_3)_2$	$Ph\text{-}C{\equiv}CH$	39†	19	3.2	16.8
$Pt_2Co_2(CO)_8(AsPh_3)_2$	$Ph\text{-}C{\equiv}CH$	40†	10	1.3	28.7
$Pt_5(CO)_6(PPh_3)_4$	$Ph\text{-}C{\equiv}CH$	100†	3	53	44
$Pt_2Co_2(CO)_8(PPh_3)_2$	$nC_6H_{13}C{\equiv}CH$	98†	55	43	—
$Pt_5(CO)_6(PPh_3)_4$	$nC_6H_{13}C{\equiv}CH$	98	10	88	—

† The i.r. spectrum (ν_{CO}) of the catalyst after the reaction is changed.

TABLE 4. HYDROGENATIONS OF DIPHENYLACETYLENE†

([Cat] $= 2.9 \pm 0.3$ mm; [Sub]/[Cat] $= 110 \pm 10$; $T = 50$ °C; pressure $= 50$ atm; time $= 20$ h; the solvent was toluene.)

catalyst	percentage conversion	percentage monohydrogenation	percentage dihydrogenation
$Pt(C_6H_{11}NC)_2[Mo(CO)_3(\eta^5\text{-}C_5H_5)]_2$	2.4	2.4‡	—
$Pt_2Co_2(CO)_8(PPh_3)_2$§	24	23‡	1
$Pt_5(CO)_6(PPh_3)_4$§	55	52.5‡	2.5

† The substrate $EtOOC\text{-}C{\equiv}C\text{-}COOEt$ is not hydrogenated; no oligomers are found in the hydrogenation of $Ph\text{-}C{\equiv}C\text{-}Ph$ under these conditions.
‡ Selectivity of more than 95 % to the *cis* isomer.
§ The i.r. spectrum (ν_{CO}) of the catalyst after the reaction is changed.

Because the only platinum product formed in this transformation is the cluster $Pt_5(CO)_6(PPh_3)_4$ (**3**), we have examined its catalytic properties under our standard conditions. The results are comparable with those found by using **1** as catalyst, but $Pt_5(CO)_6(PPh_3)_4$ can be recovered unaltered from the reaction mixture. Infrared and thin-layer investigations did not show the presence of any other complexes. Under our conditions, the decomposition of the platinum clusters to metals is negligible, but we observed extensive decomposition at higher temperatures and pressures.

Although it is very difficult to ascertain the presence of very small amounts of very dispersed colloidal metals (Laine 1982), in few experiments the formation of very limited amounts of colloidal metals was confirmed by using the Maitlis methodology (Hamlin *et al.* 1980). We did find some hydrogenation activity due to metal left on the cellulose, but the product distribution was so different from that found with the original or recycled metal complexes that the catalytic activity shown in tables 1 and 2 must be ascribed mainly to a true homogeneous process. In addition, we could not detect any hydrogenation of the aromatic rings either of styrene or of the aromatic solvents (toluene or benzene). This observation supports a true homogeneous process, because it is known that colloidal or very dispersed platinum particles are active catalysts for aromatic hydrogenation under our standard reaction conditions.

3. Hydrogenation of acetylenes

The results of the catalytic hydrogenations of terminal acetylenes ($C_6H_5C{\equiv}CH$ and $n\text{-}C_6H_{13}C{\equiv}CH$) and internal acetylene ($C_6H_5\text{-}C{\equiv}C\text{-}C_6H_5$) by catalysts **1**, **2** and **3** are given in tables 3 and 4. The internal acetylene ($COOC_2H_5\text{-}C{\equiv}C\text{-}COOC_2H_5$) was not hydrogenated by any of the clusters that we examined. All the clusters used are rather active in the hydrogenation of the terminal acetylenes. Whereas catalyst **2** is recovered unaltered from the reaction mixture, catalyst **1**, and its $AsPh_3$ homologue, and catalyst **3** transform into new species with very different infrared spectra in the ν_{CO} region.

We have so far been unable to characterize the nature of these new species; it is interesting that these new metal carbonyl complexes are also good hydrogenation catalysts for mono-olefins because they readily hydrogenate styrene. This explains why catalysts **1** and **2** are not selective catalysts for hydrogenation of terminal acetylenes to olefins, although this would be expected in view of their poor ability to hydrogenate olefins. The fast *in situ* formation of new catalytic species before hydrogenation of acetylenes to olefins is supported by the complete absence of any isomerization of 1-octene, a reaction that is easily catalysed by both original catalysts **1** and **2**.

In conclusion, we propose that new species are formed by the interaction between the original clusters and the terminal acetylenes; these new complexes are the active hydrogenation catalysts.

We have also observed during the hydrogenation of $C_6H_5C{\equiv}CH$ with catalysts **1** and **3** a secondary reaction that causes the oligomerization of phenylacetylene. Such oligomerization reactions do not occur with $n\text{-}C_6H_{13}C{\equiv}CH$, suggesting that the acidity of the acetylenic hydrogen must play a role in controlling the oligomerization process. With catalyst **2**, which has the linear $PtMo_2$ frame, the selectivity of the hydrogenation is also quite poor, but we could not detect any oligomerization with any of the terminal acetylenes investigated. Because the original complex was always recovered unaltered, the lack of selectivity is quite unexpected in view of its poor hydrogenating properties towards olefins (see table 2). We suspect that this catalyst develops, in the presence of terminal acetylenes, a specific activity for olefin hydrogenation. Catalysts **1** and **3** are active towards the hydrogenation of internal acetylenic compounds such as diphenylacetylene $C_6H_5\text{-}C{\equiv}C\text{-}C_6H_5$ (table 4). They show a remarkable and selective formation of *cis*-stilbene, together with a complete absence of any oligomerization reaction.

Catalyst **2** has poor activity, although it has a high selectivity towards *cis*-stilbene. It must be pointed out also that although catalyst **2** is recovered quantitatively from the reaction mixture catalysts **1** and **3** transform into new carbonyl species. We suppose that these new species, which are generated by the interaction of diphenylacetylene with catalysts **1** and **3**, are related to the real hydrogenation catalytic entities.

4. CONCLUSION

This investigation was initiated on the basis that the presence of coordinatively unsaturated metal atoms in a molecular metal cluster could cause high catalytic activities. However, our work does not confirm that the presence of coordinatively *open* platinum atoms in the cluster cages is related to high activities. In fact, under conditions used for olefin hydrogenation, the original cluster rearranges to new clusters that do not have coordinatively *open* platinum atoms in its metallic cage. Obviously in this case a low catalytic activity is expected. With terminal and internal acetylenes the interaction of the substrates with the original clusters produces new species, which are quite active towards the hydrogenation of acetylenes and olefins. Although we could not characterize these new active species, the disappearance in their i.r. spectra of bridging carbonyl groups suggests that mononuclear species are formed. The relatively high catalytic activities observed can be attributed to such species. In conclusion, we could not confirm our hypothesis because the original clusters are unstable under catalytic conditions.

The particular case of the high activity for the hydrogenation of terminal acetylenes by catalyst **2**, with a linear $PtMo_2$ frame, is worthy of discussion because the original cluster structure is maintained unchanged.

We could not find a parallel high activity for the hydrogenation of internal acetylenes. We therefore conclude that the slightly acidic hydrogen atom of the acetylene must be involved in the catalytic process as a 'co-catalytic' entity. This is not unexpected because it is well known that weak protonic substances may act as co-catalysts in homogeneous catalytic processes. In our case we suspect that the co-catalytic action must be related to the facile oxidative addition of the terminal acetylene to the formally platinum (II) centre to form an active platinum hydride when compared with the known difficult oxidative addition of the hydrogen molecule to platinum (II) centres (Roundhill & Jonassen 1968). This suggestion would explain the low activity displayed by catalysts **2** towards olefins and internal acetylenes (difficult activation of molecular hydrogen) compared with the high activity found with thermal acetylenes (easy formation of an intermediate active metal hydride). It is well known that the formation of active metal hydrides is necessary for a catalytic activity of hydrogenation.

REFERENCES

Barbier, J. P., Bender, R., Braunstein, P., Fischer, J. & Ricard, L. 1978 *J. chem. Res.* (*S*), pp. 230–231.
Barbier, J. P. & Braunstein, D. 1978 *J. chem. Res.* (*S*), pp. 412–413.
Basset, J. M. & Ugo, R. 1977 In *Aspects of homogeneous catalysis* (ed. R. Ugo), vol. 3, pp. 138–183. Dordrecht and Boston: D. Reidel.
Braunstein, P. & Dehand, J. 1970 *J. organometall. Chem.* **24**, 497–500.
Braunstein, P., Dehand, J. & Nennig, J. F. 1975 *J. organometall. Chem.* **92**, 117–123.
Coffey, R. S. 1970 In *Aspects of homogeneous catalysis* (ed. R. Ugo), vol. 1, pp. 16–75. Dordrecht and Boston: D. Reidel.
Dehand, J. & Nennig, J. F. 1974 *Inorg. nucl. Chem. Lett.* **10**, 875–878.
Hamlin, J. E., Hirai, K., Millan, A. & Maitlis, P. M. 1980 *J. molec. Catal.* **7**, 543–544.

Laine, R. 1982 *J. molec. Catal.* **14**, 137–169.
Norton, J. R. 1977 In *Fundamental research in homogeneous catalysis* (ed. M. Tsutsui & R. Ugo), vol. 1, pp. 99–114. New York and London: Plenum Press.
Pearson, R. G. & Dehand, J. 1969 *J. organometall. Chem.* **16**, 485–490.
Roundhill, D. M. & Jonassen, H. B. 1968 *Chem. Commun.*, pp. 1233–1234.
Smith, A. K. & Basset, J. M. 1977 *J. molec. Catal.* **2**, 229–241.

Phil. Trans. R. Soc. Lond. A **308**, 131–140 (1982)   [131]
Printed in Great Britain

Industrial aspects of cluster chemistry

By R. Whyman

*Imperial Chemical Industries p.l.c., New Science Group, P.O. Box 11, The Heath,
Runcorn, Cheshire WA7 4QE, U.K.*

The reasons for industrial interest in metal clusters and cluster catalysis are described. These may be summarized as attempting to bridge the gap between homogeneous and heterogeneous catalysis, specifically by combining the high selectivities typical of the former with the high activities associated with the latter. Progress towards the realization of this objective is illustrated by using examples of our work in both areas. Thus, some results from an investigation of homogeneous ruthenium and rhodium catalysts (as both separate components and mixtures) for the synthesis of oxygenated products from $CO–H_2$ are summarized and correlated with high-pressure infrared spectroscopic measurements. The cluster anion $[Rh_5(CO)_{15}]^-$ is shown from spectroscopic evidence to be very closely related to the catalytically active species in the rhodium-catalysed reactions. Also, the distinction between the behaviour of supported catalysts derived from Group VIII metal cluster compounds and those obtained by more conventional methods of heterogeneous catalyst preparation is discussed. For example, in the case of ruthenium, cluster-derived catalysts are shown to display greatly enhanced activity for the complete hydrogenolysis of straight-chain aliphatic hydrocarbons to methane and provide a temperature advantage of $150\,°C$ relative to conventionally prepared ruthenium catalysts, where only moderate hydrocarbon conversions are noted. The increased activity superficially correlates with the smaller metal crystallite sizes (15–20 Å; 1.5–2.0 nm) reproducibly obtainable with metal cluster compounds as catalyst precursors.

Introduction

Industrial interest in metal clusters is centred around their application as catalysts, either directly in the form of homogeneous systems or indirectly by acting as precursors to novel heterogeneous catalysts (Whyman 1980). A recent paper has described cluster catalysis as an important new research area (Haggin 1982). In that report the current situation has been accurately summarized in terms of the importance of cluster catalysis deriving more from the promise of future commercial successes than from any demonstrated use or indeed ability to solve existing problems. Most industrial research laboratories are now said to maintain major efforts in cluster research with emphasis on developing selective catalysts for C_1 chemistry.

Metal clusters are of interest as potential catalysts principally because they may be considered to represent the interface between the traditional disciplines of homogeneous and heterogeneous catalysis. Thus, on the one hand, the presence of adjacent metal sites in polynuclear complexes makes available coordinative, electronic and steric situations that cannot be duplicated at the single metal site typical of most conventional homogeneous catalysts. Catalysts derived from supported metal clusters, on the other hand, can represent one extreme, namely that of the very small particles in which almost every atom is a surface atom, of classical heterogeneous supported metal catalysts. The situation is summarized in figure 1. The areas 'cluster catalysis' and 'supported clusters' are of course very diffuse and may be further subdivided. For example, in addition to catalysis by molecular metal clusters, catalysis by colloidal metal particles should also be considered under the former heading. Molecular metal clusters may act either as catalysts in their own right *or* simply as catalyst precursors, possibly with subsequent fragmentation under reaction conditions to highly reactive species of lower nuclearity. Distinction between these two possibilities is difficult. Under the 'supported clusters' heading a distinction should be drawn between those materials in which most of the original attached ligands are retained (i.e. catalysis by supported molecular metal clusters) and those in which the ligands have been removed in an activation process (i.e. catalysis by very highly dispersed metal particles). Homogeneously catalysed processes are frequently characterized by high selectivity, whereas high activities tend to be associated with heterogeneous catalysts. It is therefore an attractive hypothesis to pursue that operation in this interfacial area between homogeneous and heterogeneous catalysis may provide a means of combining the high selectivity associated with the former with the high activity characteristic of the latter.

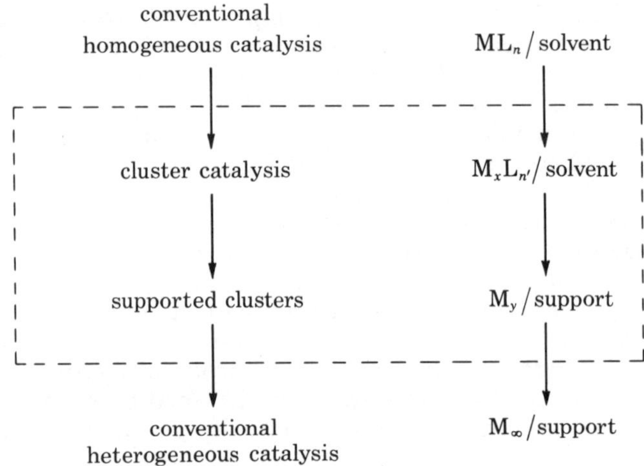

FIGURE 1. The relation between metal clusters, homogeneous and heterogeneous catalysis.

The subject of particle size effects in heterogeneous catalysis, in particular the dependence or otherwise of catalytic activity and selectivity on metal particle size in the range 1–10 nm is of considerable interest to both industrial and academic circles. General trends that have emerged suggest that transformations such as those involving the reactions of unsaturated hydrocarbons, e.g. catalytic hydrogenation and isomerization, are independent of metal particle size, i.e. are structure-insensitive. In contrast the energetically more demanding reactions such as those that involve the activation of C–H and C–C bonds in saturated hydrocarbons, and the hydrogenation of carbon monoxide and of nitrogen, are structure-sensitive. The latter are also rarely observed in the presence of conventional mononuclear homogeneous catalysts. It is conceivable that they may be facilitated by the presence of adjacent metal centres in a multinuclear cluster.

In attempting to assess whether the elegant chemistry of the type described in this Discussion Meeting can be profitably developed and applied to real industrial situations it is important to at least consider the following questions as they relate to the subject on either side of the interface. First, do homogeneous catalysts derived from metal clusters provide catalytic activities or selectivities that are not attainable by using mononuclear catalyst precursors? Secondly, do heterogeneous catalysts derived from metal clusters show genuinely different properties from supported metal catalysts prepared by traditional routes? In this paper I shall attempt to answer these questions at least in part by using two aspects of our work as illustrations. The first is related to homogeneous catalysts for the reduction of carbon monoxide and the second to reactions of saturated hydrocarbons over catalysts derived from supported metal clusters.

Homogeneously catalysed reduction of carbon monoxide

The development of catalysts for the selective synthesis of high added value oxygenated molecules such as alcohols, diols and polyols directly from $CO-H_2$ (or other C_1 molecules such as methanol) is of considerable commercial potential (see, for example, Pruett 1981) and therefore the scene of intense industrial activity. Extensive studies have been made on a rhodium-based homogeneous catalyst for the production of ethylene glycol and methanol directly from synthesis

Table 1. Conversion of $CO-H_2$ to ethylene glycol: reaction sequences observed by high-pressure i.r. spectroscopy

catalyst precursors	species present at 90 MPa and 473 K
$Rh(CO)_2acac \xrightarrow{\text{no base}} Rh_6(CO)_{16}$	$Rh_6(CO)_{16} + [Rh_5(CO)_{15}]^- \text{ (tr.)}$
$Rh(CO)_2acac \xrightarrow{\text{2-OHpy}} [Rh_5(CO)_{15}]^-$	$[Rh_5(CO)_{15}]^-$
$[Rh_{12}(CO)_{30}]^{2-} \xrightarrow{\text{no base}} [Rh_5(CO)_{15}]^- + Rh_4(CO)_{12}$	$[Rh_5(CO)_{15}]^- + Rh_6(CO)_{16}$

$$[Rh_{12}(CO)_{30}]^{2-} \xrightarrow{H_2} [HRh_6(CO)_{15}]^- \xrightarrow{CO-H_2}$$

$[Rh_{12}(CO)_{30}]^{2-} \xrightarrow{\text{2-OHpy}} [Rh_5(CO)_{15}]^-$	$[Rh_5(CO)_{15}]^-$
$[Rh_6(CO)_{15}]^{2-} \rightleftharpoons [Rh_5(CO)_{15}]^- + [Rh(CO)_4]^-$	$[Rh_5(CO)_{15}]^- + [Rh(CO)_4]^-$
$Ir(CO)_2acac \xrightarrow{\text{2-OHpy}} Ir_4(CO)_{12}$	$Ir_4(CO)_{12}$

$[Rh_{13}(CO)_{24}H_3]^{2-}$ also observed, but only under CO-deficient conditions

Table 2. Conversion of $CO-H_2$ to oxygenates by rhodium and ruthenium catalysts

(Catalyst composition: Rh added as $Rh(CO)_2acac$, Ru added as $Ru(acac)_3$; 2-hydroxypyridine (2.25 mmol), caesium benzoate (0.125 mmol) in 50 ml N-methylpyrrolidone–tetraglyme (1:4). Reaction conditions: 86.1 MPa $CO-H_2$ (1:1), 503 K, 4 h)

metal added/mmol		products formed/mmol		
Rh	Ru	CH_3OH	C_2H_5OH	$\begin{matrix} CH_2OH \\ | \\ CH_2OH \end{matrix}$
0.75	0	47.1	2.6	101.3
0.75	0.25	4.5	0.6	5.0
0.75	0.75	5.8	1.0	9.8
0	0.75	22.5	3.5	0.9

gas (Pruett & Walker 1974, 1976). High pressures (in the range 50–125 MPa) and temperatures (473–513 K) are necessary for significant conversions, the greatest selectivity to ethylene glycol being observed at higher pressures. Although both mononuclear and polynuclear rhodium complexes may be used as catalyst precursors, i.r. spectroscopic evidence has been presented to suggest the involvement of polynuclear rhodium carbonyl anions as the catalytically active species (Vidal & Walker 1980; Whyman 1980). Some of our spectroscopic results are summarized in table 1.

The use of ruthenium catalysts for the formation of methanol under similar reaction conditions has been described (Bradley 1979). No ethylene glycol is produced. In contrast to the observations on the rhodium system, the use of either polynuclear or mononuclear ruthenium complexes as catalyst precursors results in the formation of mononuclear $Ru(CO)_5$ under reaction conditions. In neither case has the nature of the active catalyst been established unequivocally.

We have studied the catalytic properties of closely related rhodium and ruthenium systems, together with their mixed-metal combinations in different proportions. In the presence of a nitrogen base and alkali-metal promoter, rhodium complexes form active catalysts for the production of ethylene glycol and methanol at 86 MPa total $CO-H_2$ pressure and 503 K (see

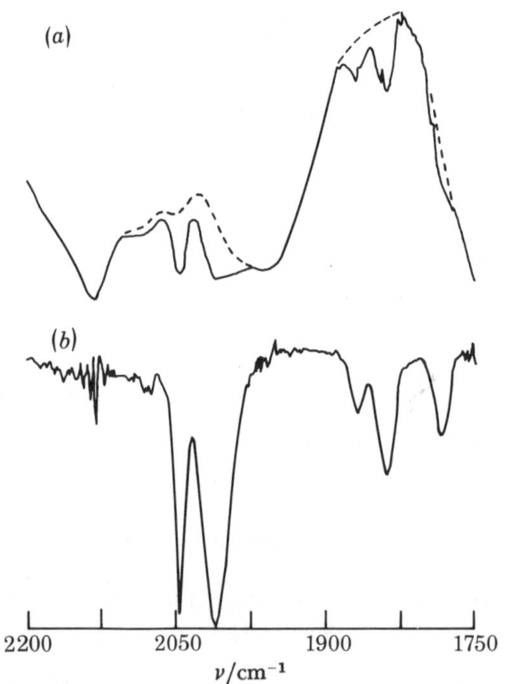

FIGURE 2. Fourier transform i.r. spectra (in the region 2200–1750 cm^{-1}) of the catalyst composition initially containing $Rh(CO)_2acac$ (0.3 mmol), 2-hydroxypyridine (0.9 mmol), caesium benzoate (0.05 mmol) in 20 ml N-methylpyrrolidone–tetraglyme (1:4) measured at 31.1 MPa $CO-H_2$ (1:1) and 323 K. (a) Spectrum of reaction mixture (——); reference spectrum of the system in the absence of rhodium complex (– – –). (b) Spectrum of the rhodium-containing species after subtraction of (– – –) from (——).

table 2). An ethylene glycol–methanol molar selectivity of 2.2:1 is observed. Analogous ruthenium catalysts are less active under these reaction conditions and produce significant quantities of methanol but only traces of ethylene glycol. The addition of up to equimolar amounts of ruthenium to the optimum rhodium catalyst for the production of ethylene glycol results in a dramatic suppression of the overall catalytic activity.

These changes in catalytic activity have been correlated with Fourier transform infrared spectroscopic measurements under similar reaction conditions and under which the major species present in solution have been identified. Spectroscopic studies on the rhodium-based catalyst are consistent with the formation of the polynuclear anion $[Rh_5(CO)_{15}]^-$ (Fumagalli et al. 1980) at 31.1 MPa $CO-H_2$ and 323 K (see figure 2), and this remains the only detectable species present in solution at 65.9 MPa and 503 K. The less complex system $Rh_4(CO)_{12}-N$-methyl-pyrrolidone–tetraglyme, which is also catalytically active for the formation of oxygenated products, shows very similar features. In the ruthenium system (see figure 3), which produces significant quantities of methanol but only minor amounts of ethylene glycol, the metal is predominantly present in the form $[HRu_3(CO)_{11}]^-$ (Johnson *et al.* 1979) at 54.0 MPa CO–H_2 and 503 K. The intermediate formation of $Ru(CO)_2(acac)_2$ ($\nu(CO)$ 2054, 1983 cm^{-1}) (Calderazzo *et al.* 1969) is observed at 423 K, but above 473 K the spectrum shows absorptions at 2010s, 1985s and 1953m cm^{-1}, which correspond closely to the spectrum of the polynuclear hydridoruthenium carbonyl anion. In the Ru–Rh mixed system $[Rh_5(CO)_{15}]^-$ is initially observed in solution at 30.4 MPa CO–H_2 and 293 K. Tris-acetylacetonatoruthenium is unreactive under these conditions

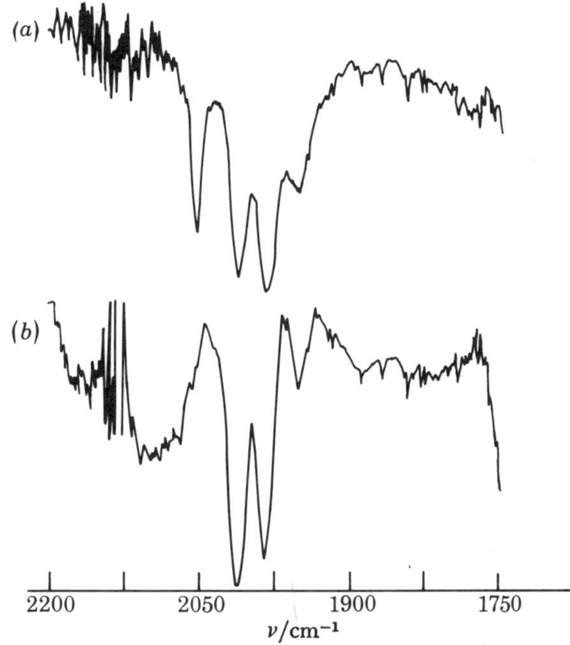

FIGURE 3. Fourier transform i.r. spectra of ruthenium-containing catalyst composition (after subtraction of background absorptions due to solvents and additives) initially comprising $Ru(acac)_3$ (0.3 mmol), 2-hydroxypyridine (0.9 mmol) and caesium benzoate (0.05 mmol) in 20 ml *N*-methylpyrrolidone–tetraglyme (1:4). (*a*) Spectrum recorded at 50.0 MPa CO–H_2 and 423 K. (*b*) Spectrum recorded at 54.0 MPa CO–H_2 and 503 K.

and is reduced to $Ru(CO)_2(acac)_2$ only when the temperature approaches 423 K. At this stage the spectrum (figure 4*a*) is consistent with the presence of a mixture of $Ru(CO)_2(acac)_2$, $[Rh_5(CO)_{15}]^-$ and a small amount of $[Rh(CO)_4]^-$ (absorption at 1896 cm^{-1}). Above this temperature the distinctive bridging $\nu(CO)$ absorptions at 1868, 1837 and 1784 cm^{-1}, characteristic of $[Rh_5(CO)_{15}]^-$, and the peak due to $[Rh(CO)_4]^-$ decrease in intensity and are replaced by a single band at 1844 cm^{-1}. At 63.5 MPa and 503 K the species $[Rh_5(CO)_{15}]^-$ is *not* present in detectable concentrations in solution. Under these conditions the infrared spectrum comprises absorption bands at 2053m, 2036sh, 2016vs, 1984s, 1953mw and 1844 cm^{-1} (see figure 4*b*). The most probable interpretation of this spectrum is in terms of the presence of a mixture of $[Rh_{13}(CO)_{24}H_3]^{2-}$ ($\nu(CO)$ 2016, 1844 cm^{-1}) (Albano *et al.* 1975) and $[HRu_3(CO)_{11}]^-$ ($\nu(CO)$ 2016, 1984, 1953 cm^{-1}) together with smaller concentrations of $Ru(CO)_2(acac)_2$ ($\nu(CO)$ 2053, 1984 cm^{-1}) and possibly $Ru(CO)_5$ ($\nu(CO)$ 2036 and 1995 cm^{-1}, the latter band being obscured). An alternative explanation could arise from the presence of at present unidentified mixed Ru–Rh species such as $[RuRh_4(CO)_{15}]^{2-}$, although the above i.r. spectrum does not resemble that reported for the related complex $[RuIr_4(CO)_{15}]^{2-}$ (Fumagalli *et al.* 1981). The apparent lack of evidence for the presence of mixed Ru–Rh species under reaction conditions is at first sight surprising. However, our knowledge of the stability of anionic mixed-metal carbonyl clusters with respect to pressures of CO or H_2, or both, is extremely limited, although fragmentation to homonuclear species has been reported to occur quite readily with neutral analogues (Fox *et al.* 1980). For example, $HCoRu_3(CO)_{13}$ in hexane reacts completely in 1 h with 0.1 MPa CO to give $Ru_3(CO)_{12}$, $Ru(CO)_5$, $Co_2(CO)_8$ and hydrogen. It is therefore quite conceivable that mixed Ru–Rh complexes would be unstable with respect to their homonuclear counterparts under severe reaction conditions.

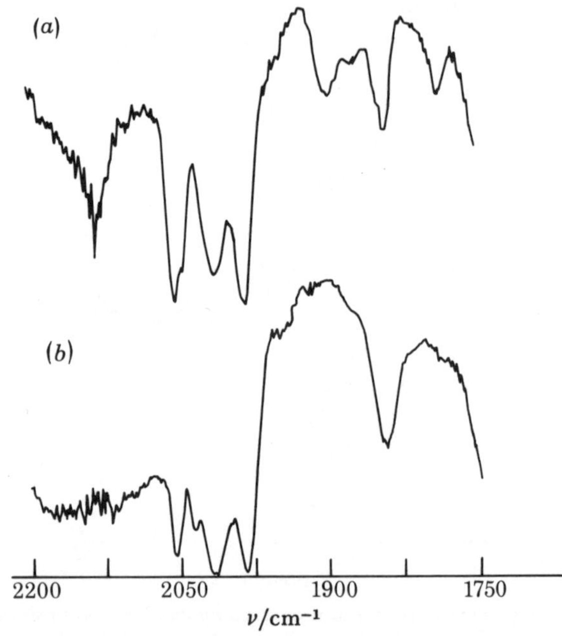

ν/cm^{-1}

FIGURE 4. Fourier transform i.r. spectra of ruthenium–rhodium catalyst composition (after subtraction of background absorption due to solvents and additives) initially comprising $Ru(acac)_3$ (0.3 mmol), $Rh(CO)_2acac$ (0.3 mmol), 2-hydroxypyridine (0.9 mmol) and caesium benzoate (0.05 mmol) in 20 ml N-methylpyrrolidone–tetraglyme (1:4). (a) Spectrum recorded at 54.0 MPa CO–H_2 and 423 K. (b) Spectrum recorded at 63.5 MPa CO–H_2 and 503 K.

TABLE 3. CONVERSION OF CO–H_2 TO OXYGENATES

(High-pressure F.t. i.r. results on Rh and Ru catalysts at 60.0 MPa and 293–503 K.)

rhodium

$Rh(CO)_2acac$–2-OHpy–CsOBz–NMP–TG

$$Rh(CO)_2acac \xrightarrow{\text{293 K}} [Rh_5(CO)_{15}]^- \xrightarrow{\text{503 K}} \text{no change}$$

ruthenium

$Ru(acac)_3$–2-OHpy–CsOBz–NMP–TG

$$Ru(acac)_3 \xrightarrow{\text{423 K}} Ru(CO)_2(acac)_2 \xrightarrow{>463\ \text{K}} [HRu_3(CO)_{11}]^-$$

ruthenium–rhodium

$$Rh(CO)_2acac \xrightarrow{\text{293 K}} [Rh_5(CO)_{15}]^- \xrightarrow{>423\ \text{K}} [Rh_{13}(CO)_{24}H_3]^{2-}$$

$$Ru(acac)_3 \xrightarrow{\text{423 K}} Ru(CO)_2(acac)_2 \xrightarrow{>463\ \text{K}} [HRu_3(CO)_{11}]^-$$

The interpretation of the spectroscopic data is summarized in table 3, and a comparison between this and the results from the catalytic studies clearly suggests that the presence of $[Rh_5(CO)_{15}]^-$ in solution is to be closely associated with the generation of catalysts forming ethylene glycol and methanol. This is not to suggest that it is the active catalyst but merely a closely related precursor. The role of $[HRu_3(CO)_{11}]^-$ is less clear, although in the monometallic system at least it acts as a precursor to a methanol-forming catalyst.

If $[Rh_5(CO)_{15}]^-$, the major species present in solution, is indeed the actual catalyst, the total activity of the system (based on rhodium concentration) is far too low for commercial application. This appears to be a general feature of homogeneous catalysts so far reported for the reduction of carbon monoxide. Under the reaction conditions described here, the addition of ruthenium to the rhodium system results in a very marked catalytic deactivation. Equally marked synergistic effects are well documented (Thompson 1975 and references therein), and the search for promotional effects (additional metal(s), ligands, bases, etc.) offers much scope for future research in this area of chemistry, irrespective of whether mononuclear, dinuclear or polynuclear metal centres are involved.

ACTIVATION OF SATURATED HYDROCARBONS OVER HETEROGENEOUS CATALYSTS DERIVED FROM SUPPORTED METAL CLUSTERS

As far as the development of heterogeneous catalysts derived from metal cluster compounds is concerned, such catalysts will be valuable only if they exhibit activities and selectivities that differ from those afforded by materials prepared conventionally. Unfortunately, it is a deficiency of much of the work so far reported that such comparisons have not been made. In an attempt to clarify this situation we initiated a programme of work to compare the catalytic activity of supported metal clusters of known particle size (prepared by the impregnation and activation–decomposition of metal cluster compounds on supports) with that displayed by catalysts containing the typical crystallite sizes available from conventional methods of heterogeneous catalyst preparation, namely halide impregnation or ion exchange, followed by hydrogen reduction. Ruthenium was selected as the initial metal for investigation, with silica as support. The precursor complexes used and eventual catalyst type are summarized in table 4.

TABLE 4. CATALYST PRECURSOR COMPLEXES

type	Ru complex
cluster	$Ru_3(CO)_{12}$
	$Ru_6C(CO)_{17}$
ion exchange	$[Ru(NH_3)_5N_2]Cl_2$
impregnation	$RuCl_3.xH_2O$

The cluster compounds were supported on silica by impregnation from organic solvents. The 'conventional' catalysts, produced by ion exchange and halide impregnation were usually supported from aqueous solutions. After drying, all materials were activated–reduced by treatment with a nitrogen–hydrogen mixture at 573–623 K for several hours. Activated catalysts were characterized by transmission electron microscopy for the measurement of metal particle size distributions both before and after catalyst testing. Typical metal particle size distributions for catalysts prepared by the different methods are summarized in table 5. From these results it is clear that the use of cluster compounds as catalyst precursors leads to the formation of particles of smaller crystallite size than those obtained from conventional methods of catalyst preparation.

However, it is equally clear that significant aggregation of the initial M_3 and M_6 cluster units has occurred during the overall supporting–activation process.

As indicated previously, the catalyst test reaction used was the activation of saturated hydrocarbons and in the results described here, specifically n-heptane. To effect this reaction a nitrogen–hydrogen–n-heptane mixture (volume ratio 100:10:1) at atmospheric pressure was passed over a silica-supported ruthenium catalyst obtained from $Ru_3(CO)_{12}$ contained in a tubular silica reactor. The temperature was increased in 50 K increments and the hydrocarbon products monitored by gas chromatography. The results (see table 6) demonstrate that n-heptane is completely hydrogenolysed to methane at 523 K, with no evidence for the formation of any other products.

TABLE 5. METAL PARTICLE SIZES (NANOMETRES) OF SILICA-SUPPORTED RUTHENIUM CATALYSTS

precursor complex	Ru (%)	particle size/nm
$Ru_3(CO)_{12}$	1.7	1.5–2.0
$Ru_6C(CO)_{17}$	1.4	1.5–2.0
$[Ru(NH_3)_5N_2]Cl_2$	2.3	2.5–3.0
$RuCl_3.xH_2O$	0.5	3.5–4.5

TABLE 6. HYDROGENOLYSIS OF n-HEPTANE OVER SILICA-SUPPORTED RUTHENIUM CATALYST

(Catalyst precursor, $Ru_3(CO)_{12}/SiO_2$; activation, N_2–H_2(4:1) at 623 K for 4 h; metal content, 1.0%; H_2:n-heptane, 10:1.)

temperature/K	heptane conversion (%)	products
373	<1	—
423	7	—
473	87	$CH_4+C_2H_6$
523	100	CH_4

TABLE 7. HYDROGENOLYSIS OF n-HEPTANE OVER SILICA-SUPPORTED RUTHENIUM CATALYSTS: CONVERSIONS AND METHANE SELECTIVITIES AS A FUNCTION OF TEMPERATURE

catalyst precursor	Ru (%)	temperature K	conversion (%)	selectivity to methane (%)
$Ru_3(CO)_{12}$	1.7	473	90.0	93.0
		523	100	100
$Ru_6C(CO)_{17}$	1.4	523	96.0	90.0
		573	100	100
$[Ru(NH_3)_5N_2]Cl_2$	2.3	523	26.0	65.0
		623	78.5	72.0
$RuCl_3.xH_2O$	0.5	523	18.0	65.0
		623	22.2	72.0

A comparison between the activities of various supported ruthenium catalysts derived from $Ru_3(CO)_{12}$, $Ru_6C(CO)_{17}$, $[Ru(NH_3)_5N_2]^{2+}$ and $RuCl_3\cdot xH_2O$ and activated–reduced under the same conditions is presented in table 7. There is clearly a very significant difference between the cluster-derived and conventional catalysts. With the former n-heptane is completely hydrogenolysed to methane at 473–523 K, whereas temperatures greater than 623 K are required to give 78 % and 25 % conversions over the catalysts based on $[Ru(NH_3)_5N_2]Cl_2$ and $RuCl_3 \cdot xH_2O$ respectively.

The new catalysts show a very high selectivity towards methane formation. In the temperature range 473–823 K no unsaturated derivatives, e.g. ethylene, are formed, and the only other product observed, in minor amounts, is ethane.

The key point to emerge from this work is that supported ruthenium catalysts derived from cluster compounds show significantly higher activity than conventionally prepared materials for the hydrogenolysis of hydrocarbons. This effect provides the opportunity to operate at as much as 150 °C below the temperature necessary to give even moderate activity over conventionally prepared catalysts. The increased activity observed with the cluster-derived catalysts bears at least a superficial relation to the smaller metal crystallite size of these materials. In addition to providing highly active ruthenium catalysts for the destructive cleavage of C–C bonds in saturated hydrocarbons, this work has led to the development of a system for the selective removal of ethylbenzene (by hydrogenolysis to toluene and methane) from aromatic C_8 fractions containing the various xylene isomers (Simpson & Whyman 1981; Whyman 1982).

Although I have shown that metal cluster-derived catalysts can display significantly different properties from their conventionally prepared counterparts, clearly considerably more detailed work is necessary to assess such features as (i) the nature of the initial interaction between the cluster compound and the support; (ii) the sequence of events that occurs on ligand removal during the catalyst activation process; (iii) the temperature at which the molecular clusters are transformed into small metal crystallites; and (iv) determination of the mildest reaction conditions under which catalytic activity can be detected.

As a start in this direction we have, jointly with workers at Hull University, studied the interactions of $Ru_3(CO)_{12}$, $Os_3(CO)_{12}$ and $Os_6(CO)_{18}$ with silica, alumina and titania. Such materials, when activated under milder reaction conditions than discussed earlier (vacuum, 523 K), give rise to a lesser degree of aggregation and the formation of clustered entities that retain some carbonyl groups. These materials are catalysts for the hydrogenolysis of ethane, the hydrogenation of ethylene and the hydrogenation of carbon monoxide and carbon dioxide. Moreover the activity of these materials is unaffected by exposure to air, in complete contrast to the behaviour of conventional ruthenium and osmium catalysts (Hunt *et al.* 1982). This further emphasizes the difference between metal cluster-derived and conventional supported metal catalysts.

CONCLUSIONS

In this paper I hope that I have been able to demonstrate that (*a*) homogeneous catalysts derived from metal clusters can display novel selectivities particularly in the area of $CO–H_2$ chemistry and that (*b*) heterogeneous catalysts derived from metal clusters – in the form of either molecular aggregates or very highly dispersed metal particles – display properties that distinguish them from conventionally prepared materials. In the former case the overall catalytic activity of such systems is currently too low for commercialization, but development of the latter area could lead to a new generation of heterogeneous catalysts, provided that novel activities and selectivities can be maintained under the more forcing conditions typical of industrial use.

I thank S. Rigby, A. F. Simpson and D. Winstanley for experimental assistance.

REFERENCES

Albano, V. G., Ceriotti, A., Chini, P., Ciani, G., Martinengo, S. & Anker, W. M. 1975 *J. chem. Soc. chem. Commun.*, pp. 859–860.

Bradley, J. S. 1979 *J. Am. chem. Soc.* **101**, 7419–7421.

Calderazzo, F., Floriani, C., Henzi, R. & L'Eplattenier, F. 1969 *J. chem. Soc.* A, pp. 1378–1386.

Fox, J. R., Gladfelter, W. L. & Geoffroy, G. L. 1980 *Inorg. Chem.* **19**, 2574–2578.

Fumagalli, A., Koetzle, J. F. & Takusagawa, F. 1981 *J. organometall. Chem.* **213**, 365–377.

Fumagalli, A., Koetzle, J. F., Takusagawa, F., Chini, P., Martinengo, S. & Heaton, B. T. 1980 *J. Am. chem. Soc.* **102**, 1740–1741.

Haggin, J. 1982 *Chem. Engng News*, 8 February, pp. 13–21.

Hunt, D. J., Jackson, S. D., Moyes, R. B., Wells, P. B. & Whyman, R. 1982 *J. chem. Soc. chem. Commun.*, pp. 85–86.

Johnson, B. F. G., Lewis, J., Raithby, P. & Suss, G. 1979 *J. chem. Soc. Dalton Trans.*, pp. 1356–1361.

Pruett, R. L. 1981 *Science, Wash.* **211**, 11–16.

Pruett, R. L. & Walker, W. E. 1974 U.S. Patent no. 3,833,634.

Pruett, R. L. & Walker, W. E. 1976 U.S. Patent no. 3,957,857.

Simpson, A. F. & Whyman, R. 1981 *J. organometall. Chem.* **213**, 157–174.

Thompson, D. T. 1975 *Platinum Metals Rev.* **19**, 88–92.

Vidal, J. L. & Walker, W. E. 1980 *Inorg. Chem.* **19**, 896–903.

Whyman, R. 1980 In *Transition metal clusters* (ed. B. F. G. Johnson), ch. 8, pp. 545–606. Chichester, New York, Brisbane and Toronto: John Wiley.

Whyman, R. 1982 U.S. Patent no. 4,331,825.

Phil. Trans. R. Soc. Lond. A **308**, 141–157 (1982)   [141]
Printed in Great Britain

Halide and chalcogenide clusters of the early transition metals

By R. E. McCarley
Ames Laboratory, U.S.D.O.E., and Department of Chemistry,
Iowa State University, Ames, Iowa 50011, U.S.A.

In the last decade remarkable progress has been made in the cluster chemistry of the early transition elements. Earlier work on the metal halide cluster compounds of Nb, Ta, Mo, W and Re served to establish the promise and potential in this area. Now cluster chemistry has been extended to halides of Sc, Y, the lanthanides, Zr and Hf, as well as to oxides, sulphides and selenides of Nb, Ta, Mo and Re. After a brief survey of significant developments in these areas, attention is focused on results from recent work on Mo and W cluster compounds. The discussion includes syntheses, structures and properties of some novel halide clusters, but major emphasis is devoted to new ternary and quaternary molybdenum oxides prepared at high temperatures. Among these solid-state materials are found discrete cluster arrays ($LiZn_2Mo_3O_8$, $Zn_3Mo_3O_8$, and $Ba_{1.14}Mo_8O_{16}$) and structures with extended metal–metal bonding ($NaMo_4O_6$, $Ba_5(Mo_4O_6)_8$, $Sc_{0.75}Zn_{1.25}Mo_4O_7$ and $Ti_{0.5}Zn_{1.5}Mo_4O_7$).

Introduction

In recent years there has been an explosive growth in our knowledge of metal–metal bonded compounds of the early transition metals, extending even into the lanthanides. This growth has come as a result of the experimental realization that reactions leading to reduction and formation of compounds with these elements in reduced oxidation states must be performed in special non-contaminating containers, often at high temperatures, and often for an extended time, e.g. weeks or months (Corbett 1981 *a*). Since reduced compounds of these elements generally react with silica at high temperature, the use of inert, sealed metal containers is dictated. In many reactions the metallic element of the compound to be formed is used as the reducing agent, e.g. as between the metal and its normal halide. The desired reaction may lack suitable gaseous or liquid phases for rapid material transport between reactants, or the surface of the metal may become blocked by the formation of an adherent film. If such conditions pertain, the desired reaction can be retarded to such a degree that thermodynamically favoured products are missed by the experimenter, and the premature conclusion reached that new reduced phases are not formed in the system. Such problems have indeed delayed progress in the development of metal cluster chemistry for particularly the lanthanide, Group III and Group IV transition elements (Corbett 1980). To some extent these problems also have plagued the realization of extensive cluster chemistry in metal chalcogenide systems of the Group V and Group VI elements. Much of the recent work reported here for molybdenum oxide compounds is illustrative of this point.

Milestone compounds (structures)

A collection of important compounds whose discovery and structure elucidation marked turning points in the further development of cluster chemistry for the early transition elements in Groups III, IV and V is given in table 1. It is remarkable that essentially all of the cluster chemistry for the lanthanide, Group III and Group IV elements has been developed within the last 10 years. From the number of compounds already known it is apparent that this area will continue to grow and mature. Because excellent discussions of the recent results in this area have been published, the systems will not be elaborated further (Corbett 1980, 1981; Simon 1981).

TABLE 1. MILESTONE CLUSTER COMPOUNDS OF THE EARLY TRANSITION
METALS AND SUBSEQUENT DISCOVERIES

periodic group	compound	cluster type	subsequent discoveries
III[a]	Sc_7Cl_{12}	M_6X_{12}	La_7I_{12}, Ce_7I_{12}, Pr_7I_{12}, Gd_7I_{12}, Er_7I_{12}, Lu_7I_{12}
	Gd_2Cl_3	$[M_4X_6]_\infty$	Y_2Cl_3, Y_2Br_3, Gd_2Br_3, Tb_2Cl_3, Tb_2Br_3, Er_2Cl_3, Tm_2Cl_3
	Sc_5Cl_8	$[MX_2 \cdot M_4X_6]_\infty$	Gd_5Br_8, Tb_5Br_8
	Sc_7Cl_{10}	$[MX_2 \cdot M_6X_8]_\infty$	Er_7I_{10}
	Er_4I_5	$[M_4X_5]_\infty$	—
	ScCl	XMMX layers	YCl, YBr, LaCl, LaBr, CeCl, CeBr, PrCl, PrBr, NdBr, GdCl, GdBr, TbCl, TbBr, DyBr, HoCl, HoBr, ErBr
IV[a]	Zr_6Cl_{15}	M_6X_{12}	Zr_6Cl_{12}, Zr_6Br_{12}, Zr_6I_{12}, $Zr_6Cl_{12} \cdot M_2ZrCl_6$ (M = Na, K, Cs)
	ZrI_2	zig-zag chain	
	ZrCl	XMMX layers	ZrBr, HfCl, $ZrClH_{0.5}$, $ZrClH_{1.0}$, $ZrBrH_{0.5}$, $ZrBrH_{1.0}$
V	Nb_3Cl_8[b]	M_3X_{13}	Nb_3Br_8[c], Nb_3I_8[c]
	$CsNb_4Cl_{11}$[d]	M_4X_{16}	$CsNb_4Br_{11}$
	$Nb_6Cl_{14} \cdot nH_2O$[e]	M_6X_{12}	Nb_6Cl_{14}[f], Nb_6F_{15}[g], Ta_6Cl_{15}[h], Ta_6Br_{15}[h], Ta_6I_{14}[i], $Mg_3Nb_6O_{11}$[j]
	NbI_{11}[k]	M_6X_8	HNb_6I_{11}, $CsNb_6I_{11}$, $CsHNb_6I_{11}$

References: [a], Corbett (1980, 1981), Simon (1981) and references therein; [b], Schafer & von Schnering (1964); [c], Simon & von Schnering (1966); [d], Broll *et al.* (1969); [e], Harned (1913); [f], Simon *et al.* (1965); [g], Schäfer *et al.* (1965); [h], Bauer & von Schnering (1968); [i], Bauer *et al.* (1965); [j], Marinder (1977); [k], Imoto & Simon (1982) and references therein.

For the transition elements of Group VI, new clusters and compounds with extended metal–metal bonding have been discovered at a startling pace in the last 10 years. Excluding the interesting variety of metal–metal bonded dimers, before 1970 the only discrete cluster compounds known were the trinuclear M_3X_{13} type first elucidated in the compound $Zn_2Mo_3O_8$ (McCarroll *et al.* 1957) and derivatives based upon the $M_6X_8^{4+}$ metal halide units (Cotton & Wilkinson 1980). Subsequently, many new cluster types have been found with both halide and chalcogenide bridging ligands, and the number of metal atoms ranging from three to twelve in the discrete units. Still more recent work has demonstrated the stability of structures with infinite chains of metal–metal bonded repeat units consisting of metal octahedra sharing edges or faces. Such compounds offer characteristics of both the elemental metals from which they are formed and the more conventional isolated cluster species. The most recent progress in these areas with emphasis on work conducted in this laboratory is summarized below according to cluster size (nuclearity).

One further note should be added here. Because space does not permit coverage of all recent work, some important compound types will not be included in this discussion. Most notable among these are the exciting and intensively studied ternary molybdenum sulphide and selenide compounds $M_xMo_6S_8$, $M_xMo_6Se_8 \cdot Mo_9Se_{11}$, $M_xMo_6S_8 \cdot Mo_{12}S_{14}$ and MMo_3S_3, often referred to as the 'Chevrel phases' (Chevrel *et al.* 1971). Several excellent discussions and reviews of these compounds have been published recently and the interested inquirer should consult these references (Yvon 1979; Fischer 1978; Corbett 1981; Burdett 1982).

Trinuclear clusters

M_3X_{13} *cluster type*

The earliest discovery of trinuclear cluster compounds was made in 1957 by McCarrol *et al.* (1957) during their examination of phases formed in the ternary molybdenum oxide system $MO-MoO_2$. This work established a series of compounds $M_2^{II}Mo_3O_8$ with M^{II} = Mg, Mn, Fe, Co, Ni, Zn or Cd, all isomorphous with the structure determined for $Zn_2Mo_3O_8$ (Ansell & Katz 1966). The structure of the triangular cluster unit shown in figure 1 consists of three MoO_6 octahedra sharing common edges such that each cluster unit is bound to 13 O atoms, hence the formulation as $Mo_3O_{13}(M_3X_{13})$ cluster type. Oxygen atoms are shared between cluster units in the hexagonal lattice as indicated in the connective formula $Zn_2Mo_3O_4O_{\frac{6}{2}}O_{\frac{3}{3}}$. Strong metal-metal bonding is indicated by diamagnetism and the short Mo–Mo distance, 2.524 (2) Å,† for the equivalent edges of the cluster units in $Zn_2Mo_3O_8$. Because each Mo^{IV} atom may contribute two electrons to the Mo–Mo cluster bonding, a total of six metal cluster electrons (m.c.es) are available to form three bonds, and thus each edge of the triangle may be considered as a Mo–Mo single bond. A subsequent m.o. treatment of the bonding by Cotton (1964) confirmed this assessment. In C_{3v} symmetry the nine d-orbitals (t_{2g} type) not used in σ-bonding to O atoms were used to construct the m.o. level ordering $1a_1^2 + 1e^4$ (bonding), $2a_1$ (essentially non-bonding) and $2e + 3e + a_2$ (antibonding). I shall return to the question of the bonding or antibonding character of the $2a_1$ orbital later.

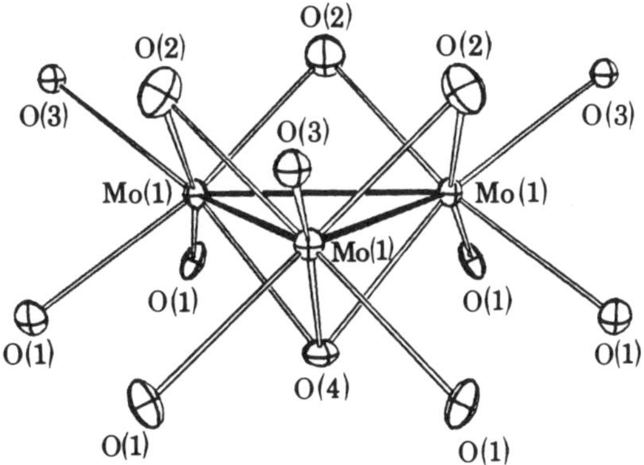

FIGURE 1. The structure of clusters of the type M_3X_{13}. The atom labelling scheme is that used in discussion of the bonding in $Zn_2Mo_3O_8$, $LiZn_2Mo_3O_8$ and $Zn_3Mo_3O_8$.

Since the initial discovery of this class of compounds, many new derivatives having the same cluster type have been characterized. These include a second solid-state series $LiM^{III}Mo_3O_8$, with M^{III} = Sc, Y (Donohue & Katz 1964) and lanthanides from Sm to Lu (Kerner-Czescleba & Tourne 1976), and many molecular complexes. Notable among the latter are $W_3O_4F_9^{5-}$, the first tritungsten cluster (Mattes & Menneman 1977); $W_3(CCH_2CMe_3)O_3Cr_3(O_2CCMe_3)_{12}$ (Katovic & McCarley), the first unit of this kind with a face-bridging alkylidyne ligand;

$$Mo_3OCl_3(O_2CMe)_3(H_2O)_3^{2+}$$

(Bino *et al.* 1979), the first such cluster with eight m.c.es; and the recent identification (Murmann & Shelton 1980) of $Mo_3O_4(H_2O)_9^{4+}$ as the red aquo-ion of Mo^{IV} previously identified as a dimer (Ardon & Pernick 1973; Cramer *et al.* 1979). In each of these cases the M–M bond distances fall

†1 Å $= 10^{-10}$ m $= 10^{-1}$ nm.

within the narrow range 2.51–2.61 Å. The Mo–Mo distance of 2.550(2) Å for the species with 8 m.c.es is especially interesting because it indicates that the addition of two electrons to the $2a_1$ m.o. has very little effect on the Mo–Mo bonding. However, the $Mo_3OCl_3(O_2CMe)_3(H_2O)_3^{2+}$ unit has three edge-bridging Cl atoms instead of the O atoms typical of other species, and their effect on Mo–Mo bond distances must also be considered.

TABLE 2. SOME IMPORTANT BOND DISTANCES (ÅNGSTRÖMS) IN
$Zn_2M_3O_8$, $LiZn_2Mo_3O_8$ AND $Zn_3Mo_3O_8$

bond†	$Zn_2Mo_3O_8$‡	$LiZn_2Mo_3O_8$	$Zn_3Mo_3O_8$
Mo(1)–Mo(1)	2.524 (2)	2.578 (1)	2.580 (2)
Mo(1)–O(1)	2.058 (10)	2.063 (6)	2.100 (9)
Mo (1)–O(2)	1.928 (20)	2.003 (8)	2.056 (13)
Mo(1)–O(3)	2.128 (30)	2.138 (5)	2.160 (8)
Mo(1)–O(4)	2.002 (30)	2.079 (7)	2.054 (11)
Mo(1)–O(av.)	2.017	2.058	2.088

† Refer to figure 1 for numbering scheme.
‡ Data from Ansell & Katz (1966).

The question of the role of the $2a_1$ m.o. in Mo_3O_{13} cluster units has been examined recently in this laboratory (Torardi & McCarley 1982). The object of the study was to prepare and determine structural parameters for compounds related to $Zn_2Mo_3O_8$, with appropriate substitution of Zn^{2+} so that reduced cluster units containing seven and eight m.c.es could be secured. Three new compounds of this type were successfully prepared at high temperatures, namely $ScZnMo_3O_8$ and $LiZn_2Mo_3O_8$, each with seven m.c.es, and $Zn_3Mo_3O_8$, with eight m.c.es, by the following reactions:

$$3Sc_2O_3 + 6ZnO + 11MoO_3 + 7Mo \xrightarrow[\text{Mo tube}]{1100\,°C} 6ScZnMo_3O_8; \tag{1}$$

$$Li_2MoO_4 + 4ZnO + 4MoO_2 + Mo \xrightarrow[\text{Mo tube}]{1100\,°C} 2LiZn_2Mo_3O_8; \tag{2}$$

$$6ZnO + 5MoO_2 + Mo \xrightarrow[\text{Mo tube}]{1100\,°C} 2Zn_3Mo_3O_8. \tag{3}$$

Single crystals resulting from (2) and (3) proved to be isomorphous and to have structures related to that of $Zn_2Mo_3O_8$, differing from the latter in the hexagonal layer stacking sequence and repeat distance. Although single crystals of $ScZnMo_3O_8$ were not obtained, X-ray powder patterns showed conclusively that this compound is isomorphous with $Zn_2Mo_3O_8$, and thus also constitutes one of the desired derivatives with seven m.c.es.

Structural parameters for $Zn_2Mo_3O_8$, $LiZn_2Mo_3O_8$ and $Zn_3Mo_3O_8$ are compared in table 2. It is seen that the principal effects of adding the seventh and eighth m.c.e. are a lengthening of the Mo–Mo bonds and a simultaneous elongation of the Mo–O bonds to the edge-bridging O atoms. Other changes in bond distances are not statistically significant within standard deviations. The observed changes with increasing m.c.e. count suggest that the seventh and eighth electrons enter a molecular orbital that is antibonding with respect to both the Mo–Mo and edge-bridging M–O interactions. Indeed, such an orbital can be formed by mixing of the one t_{2g}-type d-orbital on each metal atom not used in the M–M σ-bonding with the pπ, formally non-bonding orbitals, on the three edge-bridging O atoms. The interactions between these three d-orbitals and three oxygen pπ orbitals lead to a filled bonding set $a_1 + e$, localized mainly on the edge-bridging

O atoms, and an antibonding set $a_1^* + e^*$. In $Zn_2Mo_3O_8$, the a_1^* m.o. constitutes the l.u.m.o. and corresponds to the previously mentioned $2a_1$ level; in $LiZn_2Mo_3O_8$ and $Zn_3Mo_3O_8$ this orbital then becomes the h.o.m.o. and is occupied by one and two electrons, respectively. The magnetic properties of the latter compounds are in agreement with this scheme, indicating the filling of an orbital singlet level with the addition of one or two electrons to the $(Mo_3O_4O_{\frac{2}{2}}O_{\frac{3}{3}})^{4-}$ cluster anion. Very recent results from both electrochemical and chemical studies of the aquo ion $Mo_3O_4(H_2O)_9^{4+}$ show that chemically reversible two or three-electron reduction of the trimer can be affected, accompanied by protonation of the reduced species (Richens & Sykes 1982). The nature of the aqueous three-electron reduction product remains to be elucidated, but it seems quite likely that the two-electron reduced species is closely related, structurally and electronically, to the cluster anions of $LiZn_2Mo_3O_8$ and $Zn_3Mo_3O_8$.

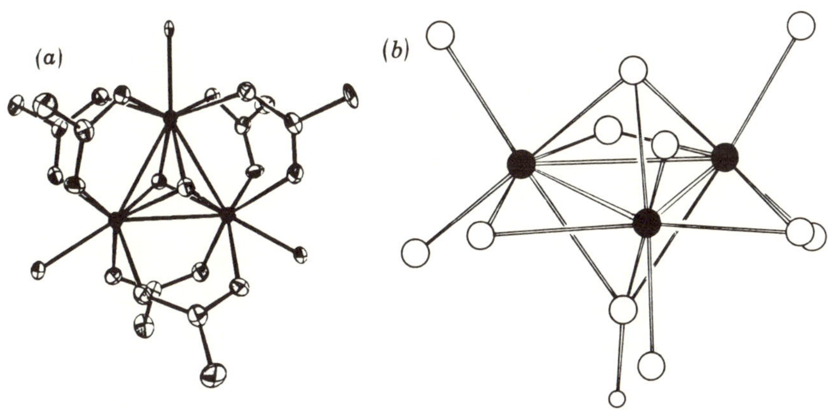

FIGURE 2. Structural representation of clusters with bicapped triangular configuration. (a) Clusters of the M_3X_{17} type; (b) clusters of the M_3X_{11} type. Metal positions are indicated by filled circles.

Bicapped M_3X_{17} and M_3X_{11} cluster types

Two additional structural types have recently been discovered for trinuclear clusters of molybdenum and tungsten. These consist of the M_3X_{17} type, exemplified by

$$W_3O_2(O_2CMe)_6(H_2O)_3^{2+} \qquad \text{(Bino } et\ al.\ 1978),$$

$$Mo_3O(CMe)(O_2CMe)_6(H_2O)_3^+ \qquad \text{(Bino } et\ al.\ 1981\,a)$$

and

$$Mo_3(CMe)_2(O_2CMe)_6(H_2O)_3^{n+} \qquad \text{(Bino } et\ al.\ 1981\,b),$$

with the triangle of metal atoms capped on either side by triply bridging O atoms, one O atom and one CCH_3, or only CCH_3 ligands, respectively. A second bicapped arrangement, of the type M_3X_{11}, has been found in the compound $MoO(ONe)_{10}$, where Ne represents the neopentyl radical (Chisholm et al. 1981). Skeletal structural arrangements in these two new cluster types are shown in figure 2.

In the M_3X_{17} cluster type two triply bridging ligands cap the face of the triangle on either side, six carboxylate ligands bridge the three edges from above and below, and three terminal aquo ligands bond in the plane of the triangle, one to each metal atom. As a result of this arrangement each metal atom is coordinated to six ligand atoms in roughly trigonal prismatic geometry with he seventh ligand bonded perpendicular to the exposed square face. Counting the metal bonds, each metal thus attains a coordination number of nine. Accordingly, the metal–metal bond distances in these derivates are longer than in the M_3X_{13} cluster type, even in those cases where both types have the same m.c.e. count and formal M–M bond order. An interesting facet of the chemistry of the M_3X_{17} clusters is that either one or two bridging μ_3-alkylidyne ligands may be incorporated into the cluster in place of the μ_3-O atoms. The clusters with two μ_3-alkylidyne ligands have been isolated in oxidized form with four or five m.c.es and average M–M bond order of $\frac{4}{6}$ or $\frac{5}{6}$, respectively. These characteristics are reflected in the M–M bond distances, which fall in the range 2.885–2.892 Å for the oxidized derivatives. A Fenske–Hall m.o. treatment of the dioxo-capped clusters of this type has recently appeared (Bursten *et al.* 1982). It was concluded that the interaction of the two capping O atoms destabilized the M–M bonding somewhat through a depopulation of the bonding a_1 m.o. constructed from the Mo $4dz^2$ atomic orbitals directed towards the centre of the cluster.

The second type of bicapped cluster was obtained as the result of addition of $MoO(ONe)_4$ across the metal–metal bond of the dimer $Mo_2(ONe)_6$:

$$MoO(ONe)_4 + Mo_2(ONe)_6 \longrightarrow Mo_3O(ONe)_{10}. \tag{4}$$

Chisholm *et al.* (1981) proposed that the addition represents the potential of a generalized method,

$$M{\equiv}M + M{\equiv}X \longrightarrow M_3X, \tag{5}$$

for the synthesis of triangulo complexes. In $Mo_3O(ONe)_{10}$ the molybdenyl O atom becomes one of the capping ligands and a neopentoxo ligand the second. The overall arrangement represents an alternative way of achieving an octahedral coordination sphere for each metal atom, as opposed to the arrangement in M_3X_{13} clusters. Thus it is not surprising that the average Mo–Mo distance of 2.529(9) Å found in $MoO(ONe)_{10}$, with six m.c.es, is closely comparable with those found in Mo_3O_{13} clusters with six m.c.es. We may expect that other examples of the M_3X_{11} cluster type will soon be found as investigations in this area continue.

TETRANUCLEAR CLUSTERS

One of the most remarkable developments in cluster chemistry has been the sudden proliferation of tetranuclear cluster units of different geometries, all occurring in compounds of Mo or W. Up to 1975 the literature provided no reports of structurally characterized Mo_4 or W_4 clusters. Subsequently the synthesis and structure of five different cluster types have been described. These are each discussed below.

Tetrahedral clusters

The solid-state compounds $GaMo_4S_8$ (Vandenberg & Brown 1975) and $Mo_4S_4Br_4$ (Perrin *et al.* 1975), formed at high temperature in reactions between the elements, each contain tetrahedral cluster units. Each structure is derived from the basic spinel structure, $A_2B_4X_8$, with Mo^{III} occupying octahedral (B) sites in the cubic close-packed anion lattice and Ga^{III} or vacancies in the tetrahedral (A) sites. The cluster units are formed by the mutual displacement of four adjacent Mo atoms from the centre of their sites towards one another to form the metal–metal bonded tetrahedron. The four faces of the tetrahedron are capped by S atoms to form a 'cubane' Mo_4S_4 unit. Each Mo is also bound to three terminal ligands, either S or Br, which are shared with other cluster units as indicated in the formulas $GaMo_4S_4S_{\frac{18}{3}}$ and $Mo_4S_4Br_{\frac{18}{3}}$. The Ga compound is electron-deficient in the sense that there are only 11 m.c.es to form the six bonds ($d(Mo–Mo) = 2.823$ Å) of the tetrahedron, but the tetrahedral cluster in $Mo_4S_4Br_4$ is bonded with the full complement of 12 m.c.es ($d(Mo–Mo) = 2.798$ Å). Although it seems quite reasonable that molecular clusters of this type might be prepared, e.g. $Mo_4S_4Br_4L_8$, none so far have been reported.

Square planar and butterfly clusters

These cluster types are grouped together because they have the common feature that they may be considered as fragments of the well known octahedral cluster units $Mo_6X_8^{4+}$. Removal of two Mo atoms from *trans* positions of the octahedron leads to the fragment Mo_4X_8 with a square arrangement of metal atoms and the eight ligands still disposed at the corners of a cube, as shown in figure 3. Chisholm *et al.* (1982) have recently prepared the Mo^{III} compound $Mo_4(OPr^i)_8Cl_4$, where the $Mo_4(OPr^i)_8^{4+}$ cluster core has the arrangement indicated above. In this case the 12 m.c.es must all reside in bonding orbitals as evidenced by the short Mo–Mo distance, $2.378(2)$ Å. This bond distance is even shorter than that expected for an average bond order of 1.5 required by 12 electrons in the square. It thus appears that the square fragment leads to unusually stable metal–metal bonding.

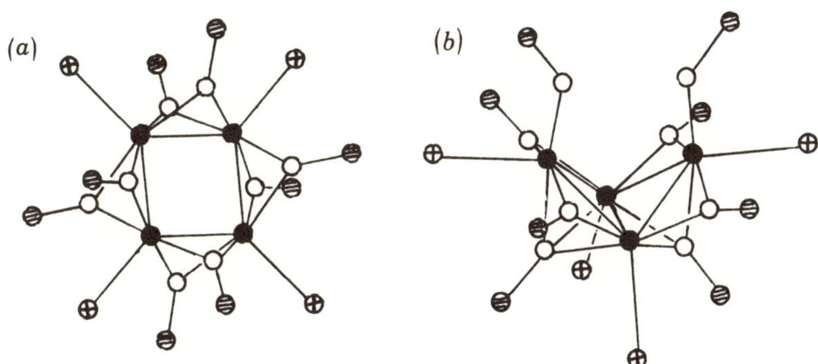

FIGURE 3. The structure of tetranuclear clusters. (*a*) The square cluster of $Mo_4(OPr^i)_8Cl_4$; (*b*) the butterfly structure of $Mo_4(OPr^i)_8Br_4$. Metal positions are indicated by filled circles, oxygen positions by open circles, carbon positions by shaded circles and halogen positions by crossed circles.

An important feature of the preparative route to $Mo_4(OPr^i)_8Cl_4$ is its formation by addition of triply bonded dimers:

$$2Mo_2(OPr^i)_6 + 4CH_3COCl \longrightarrow Mo_4(OPr^i)_8Cl_4 + 4CH_3COOPr^i. \tag{6}$$

Relatively few reactions of this type are known leading to oligomerization of multiply metal–metal bonded species. The first reactions of this type are found with quadruply bonded dimers, leading to rectangular cluster species discussed in the next section.

In the related preparation of $Mo_4(OPr^i)_8Br_4$ (Chisholm *et al.* 1982),

$$2Mo_2(OPr^i)_6 + 4CH_3COBr \longrightarrow Mo_4(OPr^i)_8Br_4 + 4CH_3COOPr^i, \tag{7}$$

which might be expected to provide the bromide analogue of the square cluster arrangement, instead leads to the butterfly cluster shown in figure 3. This butterfly or opened-tetrahedron arrangement also may be viewed as a fragment of the Mo_6X_8 octahedral cluster with two metal atoms removed from *cis* positions. The eight OPr^i ligands remain disposed roughly at the corners of a cube, although among these, two become triply bridging, four doubly bridging, and two terminal ligand types – each giving a distinct set of resonances in the ^{1}H n.m.r. spectrum. The Mo–Mo distance, 3.287 Å, corresponding to the elongated edge of the opened tetrahedron, is essentially a non-bonded distance. The five short Mo–Mo distances, 2.50 Å (averaged), again indicate very strong metal–metal bonding, a reflexion of 12 m.c.es for five bonds and an average Mo–Mo bond order of 1.20. Since the halogen atoms in these $Mo_4(OPr^i)_8X_4$ clusters are bound in terminal positions they should not play an important role in determining the geometric difference between the square and butterfly arrangements. Although the reason for this difference must be left to further investigations, we may conjecture that the kinetic pathways in (6) and (7) are different. If this is so, then the most stable cluster arrangement cannot yet be identified for these compounds.

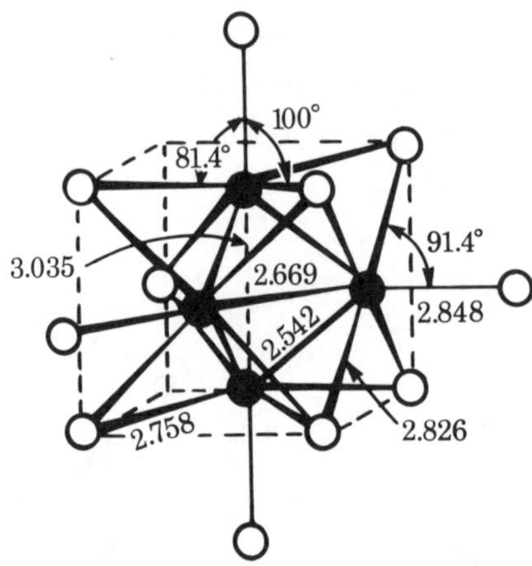

FIGURE 4. Structure of the butterfly cluster anion found in $(Bu_4N)_2Mo_4I_{11}$. Representation of the structure of Mo_4I_7 core of $Mo_4I_{11}^{2-}$ as a fragment of the $Mo_6I_8^{4+}$ cluster unit.

The first example of a butterfly four-atom fragment of the Mo_6X_8 octahedral clusters was found in the compound $(Bu_4N)_2Mo_4I_{11}$ (Stensvad *et al.* 1978). In this case one of the eight ligands at the corners of the cube is lost. The seven remaining ligands assume the positions shown in figure 4, and are divided into two triply bridging and five doubly bridging modes. Each metal bears one terminal ligand to complete the coordination sphere. A clear distinction of the three Mo–I bonding modes is found in the I $3d_{\frac{5}{2},\frac{3}{2}}$ X-ray photoelectron spectrum, which can be deconvoluted to provide three bands in the intensity ratio 2:5:4 expected for binding energies in the order $I^{tb} > I^{db} > I^t$. The $Mo_4I_{11}^{2-}$ cluster presents a greater problem for the interpretation of bonding because 15 m.c.es may participate, and also because the opened edge of the tetrahedron, $d(Mo–Mo) = 3.035$ Å, is not clearly a non-bonded distance. A rough m.o. analysis of bonding has led to the conclusion that 12 electrons reside in bonding m.o.'s and the three additional electrons occupy essentially non-bonding orbitals localized on the Mo atoms at the ends of the long bond. This analysis of the bonding is consistent with the structures of both $Mo_4(OPr^i)_8Br_4$ and $Mo_4I_{11}^{2-}$ which have comparable Mo–Mo bond distances for the five short bonds in the butterfly units. Addition of three non-bonding electrons in $Mo_4I_{11}^{2-}$ hardly disturbs the structure. On this basis a series of compounds having this structure type might be expected with m.c.e. counts ranging between 12 and 16.

Rectangular clusters

These clusters were first discovered (McGinnis *et al.* 1978) as the result of addition of the
'activated' quadruply bonded dimers, $Mo_2Cl_4(PPh_3)_2(MeOH)_2$, with ligand elimination as
follows:

$$2Mo_2Cl_4(PPh_3)_2(MeOH)_2 \xrightarrow[25\,°C]{C_6H_6} Mo_4Cl_8(PPh_3)_4 + 4MeOH; \tag{8}$$

$$2Mo_2Cl_4(PPh_3)_2(MeOH)_2 \xrightarrow[HCl.\ 60\,°C]{CH_3OH-C_{10}H_{12}} Mo_4Cl_8(MeOH)_4 + 4PPh_3. \tag{9}$$

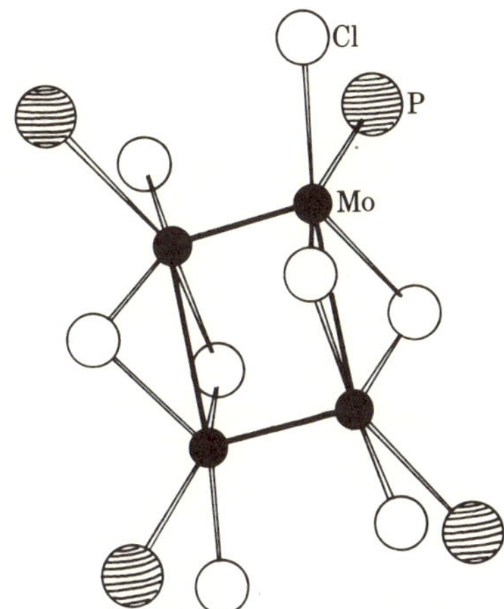

FIGURE 5. The structure of the rectangular cluster unit of $Mo_4Cl_8(PEt_3)_4$. For clarity the
ethyl groups of the PEt_3 ligands have been omitted.

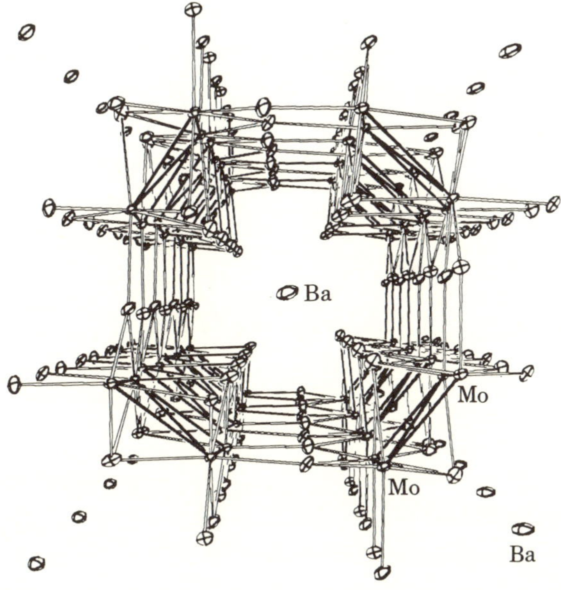

FIGURE 6. A three-dimensional view of the structure of $Ba_{1.14}Mo_8O_{16}$ as seen down the c-axis with
the eye directly in line with the central stack of Ba atoms.

Subsequent replacement of PPh_3 or MeOH with PEt_3 or PBu_3^n led to crystalline derivatives whose structure showed the presence of rectangular clusters with the basic framework shown in figure 5. Bond distances within the Mo_4 rectangle of $Mo_4Cl_8(PEt_3)_4$, 2.211(3) and 2.901(2) Å, indicate that the short and long distances correspond to Mo–Mo triple bonds and single bonds, respectively. Schematically the addition can be represented as

$$\begin{array}{ccc} M\equiv M & & M\equiv M \\ + & \longrightarrow & | \quad | \\ M\equiv M & & M\equiv M \end{array} \tag{10}$$

Studies of the electronic spectral changes that take place on going from dimers to the rectangular clusters indicate that the δ-bonding components of the quadruple bonds in the dimers are recast to form the σ-bonds corresponding to the long edges of the rectangle (Ryan & McCarley 1982). This interesting bond arrangement is unprecedented in chemistry; its analogue among organic compounds, cyclobutadiyne, is unlikely to be stable because of bond angle strain.

Subsequent synthetic and structural work has shown that the formation of the rectangular clusters can be efficiently achieved by a variety of approaches, and for tungsten as well as molybdenum (McCarley et al. 1981):

$$2K_4Mo_2Cl_8 + 4PEt_3 \xrightarrow[\text{reflux}]{\text{CH}_3\text{OH}} Mo_4Cl_8(PEt_3)_4 + 8KCl; \tag{11}$$

$$2Mo_2(O_2CMe)_4 + 4AlCl_3 + 4PEt_3 \xrightarrow[\text{25 °C}]{\text{THF}} Mo_4Cl_8(PEt_3)_4 + 4AlCl(O_2CMe)_2; \tag{12}$$

$$2Mo_2Cl_4(PBu_3^n)_4 + 2Mo(CO)_6 \xrightarrow[\text{reflux}]{\text{C}_6\text{H}_5\text{Cl}} Mo_4Cl_8(PBu_3^n)_4 + 2Mo(CO)_4(PBu_3^n)_2 + 4CO; \tag{13}$$

$$2W_2Cl_6(THF)_2(PBu_3^n)_2 + 4Na(Hg) \xrightarrow{\text{THF}} W_4Cl_8(PBu_3^n)_4 + 4NaCl + 4THF. \tag{14}$$

By using the approach represented in (12) it is possible to form the interesting mixed-metal cluster $Mo_2W_2Cl_8(PMe_3)_4$, via

$$2MoW(O_2CCMe_3)_4 + 4PMe_3 + 8AlCl_3 \xrightarrow[\text{reflux}]{\text{THF}} Mo_2W_2Cl_8(PMe_3)_4 + 8AlCl_2(O_2CCMe_3) \tag{15}$$

(Carlin & McCarley). A structural determination of this compound indeed shows the presence of the rectangular cluster, but because of disordering of the metal positions, the particular isomer formed could not be identified. From the mode of formation, starting with the pure heteronuclear dimer $MoW(O_2CCMe_3)_4$, the expected product should consist of one of the following isomers,

$$\begin{array}{ccccc} Mo\equiv W & & & Mo\equiv W & \\ | \quad | & & \text{or} & | \quad | & , \\ Mo\equiv W & & & W\equiv Mo & \end{array}$$

or a mixture of them. It is hoped that a study of the ^{31}P n.m.r. spectra, currently in progress, will permit a definitive answer to this problem.

Rhomboidal clusters

As a result of recent work in this laboratory on reduced ternary oxides of molybdenum, a new family of strongly metal–metal bonded compounds has been discovered (McCarley et al. 1981; Torardi & McCarley 1981). One member of this family is the interesting and puzzling compound $Ba_{1.14}Mo_8O_{16}$, prepared at 1100 °C via

$$1.14BaMoO_4 + 5.72MoO_2 + 1.14Mo \longrightarrow Ba_{1.14}Mo_8O_{16}. \tag{16}$$

The structure of this compound proved to be a low-symmetry variant (triclinic, $P\bar{1}$) of the tetragonal hollandite structure, e.g. $BaMn_8O_{16}$. A view of the structure looking down the unique tunnel axis, in this case the c-axis, is shown in figure 6. The structural framework is built up from infinite chains of rhomboidal cluster units running parallel to the c-axis. The chains are then woven together in the indicated square pattern by $Mo-O-Mo$ cross-linking to form the tunnels in which the Ba^{2+} cations are located. A view of one of the individual cluster chains is shown in figure 7. From the mode of interlinking of cluster units along the chains and sharing of O atoms between chains, the connectivity formula for the individual cluster units can be written as $Mo_4O_2O_{\frac{8}{2}}O_{\frac{6}{3}} = Mo_4O_8$.

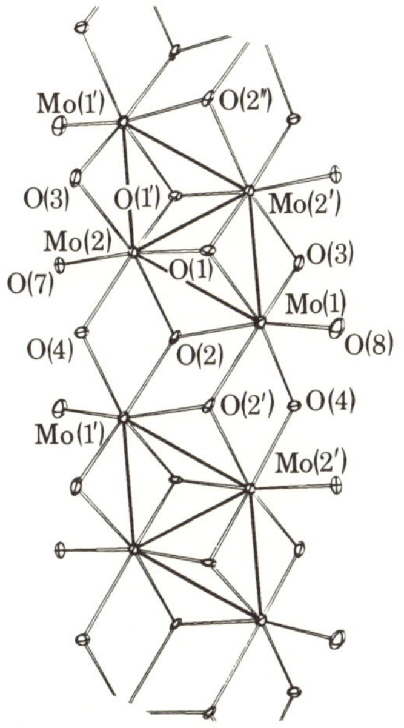

FIGURE 7. View of a segment of one infinite chain in the structure of $Ba_{1.14}Mo_8O_{16}$, showing the discrete rhomboidal cluster units. The atom labelling scheme is that used in the text.

In the $P\bar{1}$ space group the infinite chains are related in pairs. Thus, of the four chains forming the sides of each tunnel, there are two non-equivalent pairs. These non-equivalent pairs contain non-equivalent rhomboidal clusters, which will be referred to as regular and distorted units. With reference to figure 7 the $Mo(1)-Mo(2)$, $Mo(1)-Mo(2')$ and $Mo(2)-Mo(2')$ distances are, respectively, 2.616(1), 2.578(1) and 2.578(1) Å for the regular cluster units, and 2.847(1), 2.546(1) and 2.560(1) Å for the distorted units. We see that the distortion results from the elongation of two sides of the rhomboid from 2.616 Å in the regular units to 2.847 Å in the distorted units, whereas the dimensions of the other bonds of the two rhomboids remain comparable. The change in bond order associated with this bond lengthening is approximately from 1.0 to 0.5, which suggests that the $Mo(1)-Mo(2)$ bonds in the distorted units are one-electron bonds. It all other $Mo-Mo$ bonds are taken as normal two-electron bonds, then the total m.c.e. count in the regular clusters is ten, and that in the distorted clusters is approximately eight. Based on these considerations the compound may be reformulated as $Ba_{1.14}(Mo_4O_8^{2-})(Mo_4O_8^{0.28-})$.

Although the rhomboidal units are the first of their kind found in oxide systems, regular, ten-electron clusters are strictly analogous to the rhomboidal cluster units found in the halide compounds $M^INb_4X_{11}$ ($M^I = Rb, Cs; X = Cl, Br$) (Broll *et al.* 1969). Moreover, almost simultaneously with our discovery of $Ba_{1.14}Mo_8O_{16}$, the compound $W_4(OEt)_{16}$ was discovered by Chisholm *et al.* (1981 *b*). The latter molecular cluster also contains eight m.c.es and is distorted in exactly the same way as the $Mo_4O_8^{0.28-}$ units in $Ba_{1.14}Mo_8O_{16}$. Cotton & Fang (1982) have ascribed the observed distortion of these clusters to a second-order Jahn–Teller effect. This seems quite plausible for the distortion in the molecular $W_4(OEt)_{16}$, but is questionable for the distorted units in $Ba_{1.14}Mo_8O_{16}$, where other solid-state effects complicate the interpretation. In particular the $Mo(1)$–$O(8)$ bond contracts from $2.022(6)$ Å in the regular cluster to $1.894(6)$ Å in the distorted cluster. Evidently the atomic orbital utilized on $Mo(1)$ to form the $Mo(1)$–$Mo(2)$ bond becomes more strongly involved as a π-acceptor orbital towards $O(8)$ when the m.c.e. count is lowered from ten to eight.

Future work will doubtlessly uncover other examples of structures of this type. Indeed, it may be possible to design compounds with a range of m.c.e. counts from eight to ten or greater in the rhomboidal units by the proper selection of cations incorporated into the channels and the replacement of molybdenum with transition elements of differing valence-electron population. Although in $Ba_{1.14}Mo_8O_{16}$ the regular and distorted units appear to be incorporated in separate chains, other arrangements can be envisaged where varying patterns of both types are included in the same chain. Some evidence for such arrangements has been obtained from the structures of mixed-cation compounds $Li_xBaMo_8O_{16}$ and $Na_xBaMo_8O_{16}$, with $x < 1$ (McCarley & Lii 1982).

INFINITE-CHAIN COMPOUNDS

The discovery of this family of compounds was quite unintentional and resulted from a most innocent experiment. Having determined that the Li^+ and Zn^{2+} ions in $LiZn_2Mo_3O_8$ were scrambled among themselves in both octahedral and tetrahedral sites, attempts were planned for preparation of the Na^+-substituted isomorph $NaZn_2Mo_3O_8$. It was expected that Na^+ could only occupy octahedral sites and hence would induce an ordered arrangement in the cation positions. Since sodium salts commonly cause devitrification of fused silica, the mixture containing Na_2MoO_4, ZnO, MoO_2 and Mo was sealed in a Mo tube and the reaction conducted at $1100\,°C$ for several days. Upon opening the Mo tube, thin air-stable needles with silvery lustre were found growing from the walls of the tube. The structural determination of one of these needles indicated that the actual composition was $NaMo_4O_6$, and electron microprobe analysis confirmed the absence of zinc in the crystals. A subsequent reaction,

$$Na_2MoO_4 + 4MoO_2 + 3Mo \xrightarrow[1100\,°C]{Mo\ tube} 2NaMo_4O_6, \tag{17}$$

provided essentially pure, highly crystalline $NaMo_4O_6$.

The structure of this compound was astonishing and unprecedented among metal oxide systems (Torardi & McCarley 1979). A view of the structure looking down the c-axis of the tetragonal crystal system ($a = 9.570(3)$, $c = 2.8634(8)$ Å) is shown in figure 8. This view shows infinite chains composed of Mo_4O_6 repeat units interlinked through Mo–O–Mo bonds to form the square pattern and tunnels in which the Na^+ ions are located. A detailed view of a segment of one of the infinite chains, shown in figure 9, reveals that the chain consists of condensed octahedral cluster units Mo_6O_{12} fused on opposite edges by sharing Mo and O atoms with adjacent units. Including the Mo–O bonds to atoms of neighbouring chains the connective formula $Na[(Mo_2Mo_{4/2}O_{8/2}O_{2/2})O_{2/2}^a]$ is derived. It is undoubtedly the strong metal–metal bonding along the chains that dictates this structure. The metal–metal distances parallel to the chain axis for both the crystallographically independent apex and waist Mo atoms of the octahedral units are equal to the c lattice dimension, $2.8618(2)$ Å, as required by the space group $P4/mbm$. All other Mo–Mo distances within the octahedral units are much shorter; these are enumerated in table 3 for comparison with other structures.

Because of the infinite metal–metal bonded chains parallel to the c-axis, with much larger metal–metal spacings between chains, $NaMo_4O_6$ is expected to behave as a highly anisotropic, nearly one-dimensional conductor. A simple two-probe resistivity measurement along the c-axis of a single crystal provided a value of $ca.$ 10^{-4} $\Omega\,cm$, indicative of metallic behaviour. A relatively

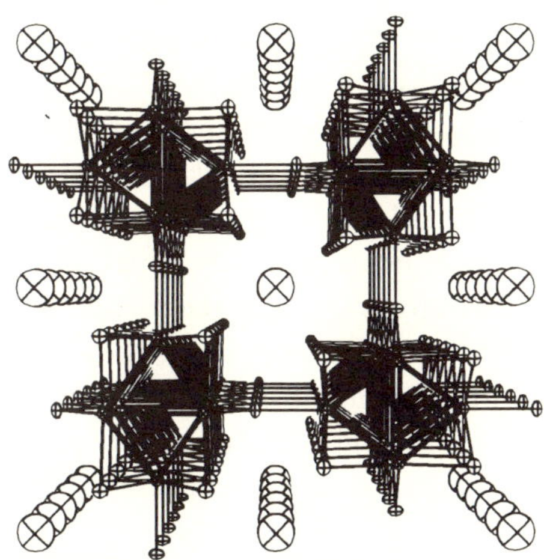

FIGURE 8. A three-dimensional representation of the structure of $NaMo_4O_6$ as viewed down the tetragonal c-axis with the eye directly over the central stack of Na atoms.

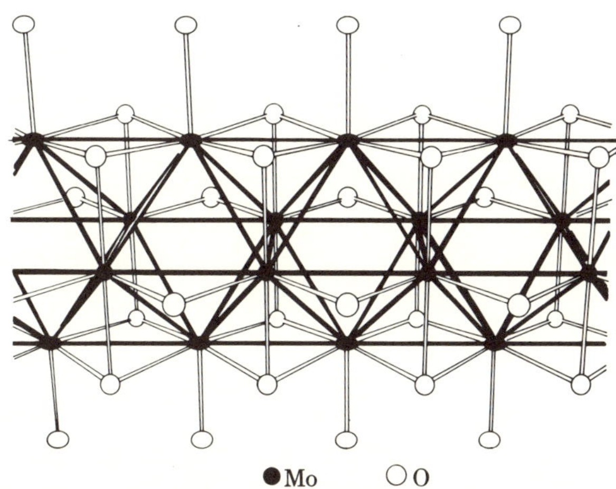

FIGURE 9. Structure of a segment of one infinite chain containing the octahedral cluster units fused on opposite edges by sharing of Mo atoms between repeat units in the compound $NaMo_4O_6$. Mo–Mo bonds are represented by heavy filled lines and Mo–O bonds by unfilled lines.

TABLE 3. A COMPARISON OF Mo–Mo BOND DISTANCES (ÅNGSTRÖMS) IN COMPOUNDS WITH INFINITE CHAIN STRUCTURES

bond type	$NaMo_4O_6$	$Sc_{0.75}Zn_{1.25}Mo_4O_7$	$Ti_{0.5}Zn_{1.5}Mo_4O_7$
apex–apex	2.8618 (2)	2.623 (2)	2.790 (2)
		3.140 (2)	2.992 (2)
apex–waist	2.780 (2)	2.749 (1)	2.741 (1)
		2.782 (1)	2.757 (1)
(waist–waist)$_\parallel$	2.8618 (2)	2.8857 (5)	2.8928 (5)
(waist–waist)$_\perp$	2.753 (3)	2.818 (2)	2.785 (2)

high density of states at the Fermi level as determined from ultraviolet photoelectron spectroscopic measurements on freshly prepared and carefully handled material also confirms the metallic character. However, measurements on pressed and sintered pellets over the range 2–300 K provide much higher resistivity values, e.g. *ca.* $10^{-2}\,\Omega$ cm at 298 K, which decrease with increasing temperature, and show a maximum at *ca.* 11 K. The higher resistivity for a polycrystalline sample is consistent with the expected anisotropy, but the temperature dependence is not understood.

In $NaMo_4O_6$ there are 13 m.c.es per $Mo_4O_6^-$ repeat unit. That all 13 m.c.es enter into bonding interactions between the Mo atoms is confirmed by the Pauling bond order sum (Corbett 1981 *c*) for the 13 net Mo–Mo bonds within each octahedral unit. The expected bond order sum for 13 electrons and 13 bonds is 6.50, and that derived from summing bond orders computed from the individual Mo–Mo distances is 6.36. We may then ask: To what extent is the electron count for such a chain structure variable? What is the total range of electrons per repeat unit possible in stable structures, and what is the limiting number of electrons in bonding states? To some extent the question of stability of the chain structure is related to the matching between van der Waals radii of the non-metallic and metallic bond radii of the metallic constituents. This is a consequence of the requirement that the average metal–metal and non-metal–non-metal spacings along the chain axis must be equal. A large mismatch of these radii is likely to destabilize the structure relative to another with greater bonding and packing efficiency. Thus, metals with large metallic radii and low valence electron count will require a large M–M distance (low net bond order) matched by relatively large anions for equal anion–anion spacing. Chains constructed from such constituents will have low m.c.e. counts. Compounds with such chain structures meeting these conditions were the first discovered, e.g. $Gd_2Cl_3(Gd_4Cl_6)$, $ScCl_2^+Sc_4Cl_6^-$ (Sc_5Cl_8) and others listed in table 1. So far the lower limit of m.c.e. per repeat unit established from known compounds is six, as in the structural variants represented by Gd_2Cl_3 and Tb_2Br_3. The anion chain $Sc_4Cl_6^-$ found in Sc_5Cl_8 is slightly more tightly bound, with seven m.c.es per repeat unit.

If the lower limit of m.c.e.s per repeat unit is about six, then what is the upper limit? Is it possible to go beyond the 13 m.c.es of $NaMo_4O_6$? Subsequent synthetic and structural work in this laboratory has answered this question in the affirmative. Experiments designed to prepare new $M_xMo_4O_6$ chain structures with M = di-, tri- or tetra-positive cations are in progress. Early success has been found with the preparation and structure determination of $Ba_5(Mo_4O_6)_8$, $Sc_{0.75}Zn_{1.25}Mo_4O_7$ and $Ti_{0.5}Zn_{1.50}Mo_4O_7$. The structure of $Ba_5(Mo_4O_6)_5$ is a low-symmetry (orthorhombic) version of that of $NaMo_4O_6$ (Torardi & McCarley 1981). A small net contraction of the average Mo–Mo bond distance in the repeat units of the chains indicates that all 13.25 m.c.es in the repeat units of $Ba_5(Mo_4O_6)_8$ do indeed reside in bonding states.

The discovery of $Sc_{0.75}Zn_{1.25}Mo_4O_7$ and $Ti_{0.5}Zn_{1.50}Mo_4O_7$ was again serendipitous (Brough *et al.* 1982). During attempts to grow single crystals of $ScZnMo_3O_8$ (see above), this material was subjected to heating at increasing temperatures. At *ca.* 1450 °C, $ScZnMo_3O_8$ decomposed or underwent reaction with the molybdenum container and beautiful gem-like crystals of the new product were formed. Likewise, $Ti_{0.5}Zn_{1.5}Mo_4O_7$ was formed at *ca.* 1450 °C in a reaction between TiO_2, ZnO, MoO_2 and Mo held in a Mo tube for several days. The composition of the new compounds was established from electron microprobe analyses of single crystals and the subsequent X-ray structure determination. Both compounds crystallize with orthorhombic unit cells in the space group Imam. Although the compounds are isomorphous there are some significant differences in Mo–Mo bond distances, signifying differences in m.c.es per repeat unit as discussed below.

Some important structural features of $Sc_{0.75}Zn_{1.25}Mo_4O_7$ are evident from a view parallel to the c-axis shown in figure 10. This view illustrates the octahedral cluster units $Mo_4O_7^{n-}$ of infinite chains running parallel to the c-axis. The chains are interlinked in the ac plane by Mo–O–Mo bridge bonding, and in the b-direction by Mo–O–Sc and Mo–O–Zn bonding. If the infinite chains can be regarded as cylinders of infinite length the overall structure can be regarded as a closest-stacking of these cylinders with Sc^{3+} and Zn^{2+} located in sites along the narrow channels created by the cylinder stacking. The sites for these cations are of two types, namely tetrahedral sites occupied only by Zn^{2+}, and octahedral sites occupied by either $Zn^{2+}(25\,\%)$ or $Sc^{3+}(75\,\%)$.

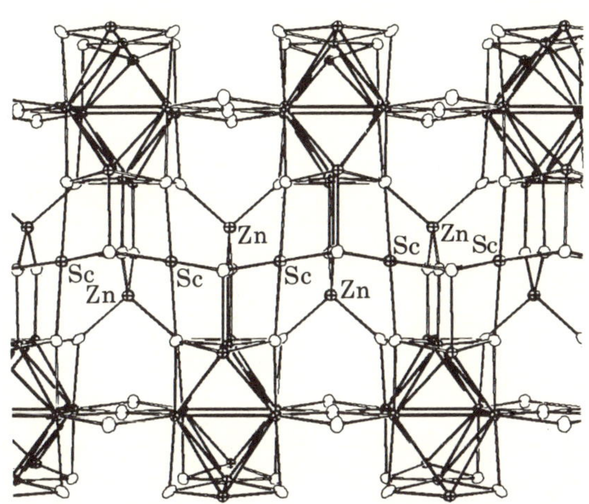

Figure 10. A three-dimensional representation of the structure of $Sc_{0.75}Zn_{1.25}Mo_4O_7$ as viewed down the ortho-rhombic c-axis. The positions of Sc and Zn atoms are labelled. The Mo_4O_7 cluster units show the Mo–Mo bonds with heavy filled lines.

Within the infinite chains the sharing of Mo and O atoms between adjacent units is much like that found in $NaMo_4O_6$. However, an extra O atom is inserted in the terminal positions on the apex Mo atoms along the chain for bonding to the Zn and Sc atoms. Because of alternate short and long Mo–Mo distances between apex atoms along the chains, the true repeat unit contains two individual cluster units. These considerations are then represented by the connective formula $Sc_{1.5}Zn_{2.5}[(Mo_6Mo_{\frac{4}{2}}O_4O_{\frac{4}{2}}O_{\frac{2}{2}})O_4]$. Assuming complete electron transfer from Sc and Zn to the cluster chains the m.c.e. count in the individual octahedral units becomes 14.75 for $Sc_{0.75}Zn_{1.25}Mo_4O_7$.

One effect of the increased m.c.e. count can be seen in the view of the cluster chains perpendicular to the chain direction, as shown in figure 11. The alternate short and long distances between apex Mo atoms is clearly evident, whereas the distances between waist Mo atoms remains perfectly regular. A comparison of Mo–Mo distances for $NaMo_4O_6$, $Sc_{0.75}Zn_{1.25}Mo_4O_7$ and $Ti_{0.5}Zn_{1.5}Mo_4O_7$ is given in table 3. The most notable feature of this comparison is that the distortion of the apex–apex bonding is quite severe in $Sc_{0.75}Zn_{1.25}Mo_4O_7$. In fact the short apex–apex distance agrees well with that expected for a Mo–Mo single bond, and the long distance is so great that it must be essentially non-bonding. A similar but less severe distortion is evident for the titanium compound. Again, Pauling bond-order values are instructive in the comparison between these structures. However, the titanium compound presents a special problem because the oxidation state of the Ti atom in this structure is not clear, and hence the number of electrons transferred to the anion chains is uncertain. For the scandium compound it is reasonably certain that this element is present as Sc^{3+} and the bond-order sum computed from the Mo–Mo distances for the individual cluster units is 6.74, a value greater by 0.38 units than that of $NaMo_4O_6$. It thus appears that additional electrons have been fed into bonding states. The increase results predominantly from the formation of the short apex–apex bonds. This is demonstrated by comparing the bond-order sum for the short and long apex–apex distances with that of two waist–waist bonds in the scandium compound. The derived values for the apex–apex and waist–waist sums are 1.10 and 0.70, respectively. The difference of 0.40 shows that the net Mo–Mo bonding is strengthened by the distortion and accommodates additional electrons in the structure. Similar considerations for the titanium compound indicate that fewer electrons are donated to the anion chains. This in turn suggests that Ti is present as Ti^{3+} and not as Ti^{4+}.

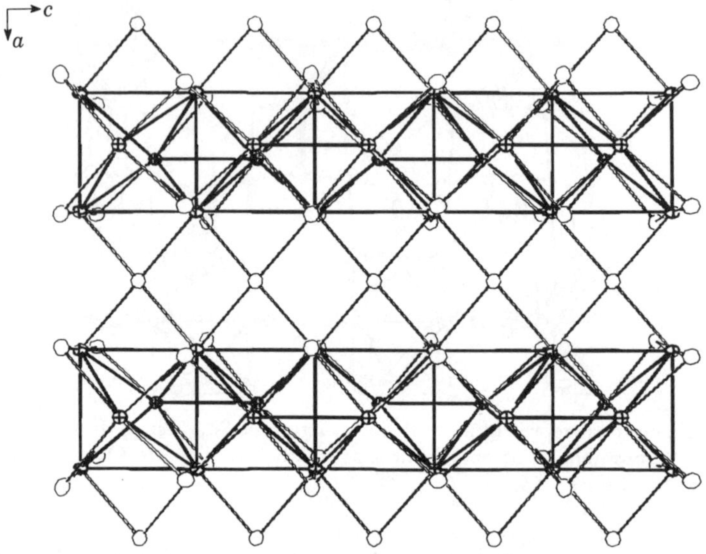

FIGURE 11. A view of the infinite chains containing octahedral cluster units fused on opposite edges in the compound $Sc_{0.75}Zn_{1.25}Mo_4O_7$. In this view the alternate short and long Mo–Mo distances between apex Mo atoms of the octahedral cluster units can be seen and the Mo–O–Mo bridge bonding connecting one chain to its neighbours also is illustrated.

I conclude by noting that this venture into the study of reduced ternary and quaternary molybdenum oxides has been most surprising and rewarding. In retrospect it is astonishing that before this work structurally characterized oxide compounds containing molybdenum in an oxidation state less than $4+$ were unknown. The potential for further interesting compounds in this class is considerable, not only for molybdenum but for other heavy transition metals as well

I give special thanks to all my coworkers, past and present, who contributed to the work reported here, and without whom I would have had little to write. Their contributions are noted in the references. Thanks also go to the U.S. Department of Energy, Basic Energy Sciences, for their continuing support of this work through the Ames Laboratory, which is operated for the U.S. Department of Energy by Iowa State University under contract no. W-7405-Eng-82. This research was supported by the Assistant Secretary for Energy Research, Office of Basic Energy Sciences, no. WPAS-KC-02-03.

REFERENCES

Ansell, G. B. & Katz, L. 1966 *Acta crystallogr.* **21**, 482.
Ardon, M. & Pernick, A. 1973 *J. Am. chem. Soc.*, **95**, 6871–6872.
Bauer, D., von Schnering, H.-G. & Schäfer, H. 1965 *J. less-common Metals*, **8**, 388–401.
Bauer, D. & von Schnering, H.-G. 1968 *Z. anorg. allg. Chem.* **361**, 259–276.
Bino, A., Cotton, F. A., Dori, Z., Koch, S., Küppers, H., Millar, M. & Sekutowsky, J. C. 1978 *Inorg. Chem.* **17**, 3245–3253.
Bino, A. Cotton, F. A. & Dori, Z. 1979 *Inorg. chim. Acta* **33**, 133–134.
Bino, A., Cotton, F. A., Dori, Z. & Kolthammer, W. S. 1981 *a J. Am. chem. Soc.* **103**, 5779–5784.
Bino, A., Cotton, F. A. & Dori, Z. 1981 *b J. Am. chem. Soc.* **103**, 243–244.
Broll, A., Simon, A., von Schnering, H.-G. & Schäfer, H. 1969 *Z. anorg. allg. Chem.* **367**, 1–18.
Brough, L. F., Carlin, R. T. & McCarley, R. E. 1982 (In preparation.)
Burdett, J. K. & Lin, J.-H. 1982 *Inorg. Chem.* **21**, 5–10.
Bursten, B. E., Cotton, F. A., Hall, M. B. & Majjar, R. C. 1982 *Inorg. Chem.* **21**, 302–307.
Carlin, R. T. & McCarley, R. E. 1982 (In preparation.)
Chevrel, R., Sergent, M. & Prigent, J. 1971 *J. solid State Chem.* **3**, 315.
Chisholm, M. H., Folting, K., Huffman, J. C. & Kirkpatrick, C. C. 1981 *J. Am. chem. Soc.* **103**, 5967–5968.
Chisholm, M. H., Huffman, J. C., Kirkpatrick, C. C., Leonelli, J. & Folting, K. 1981 *J. Am. chem. Soc.* **103**, 6093–6099.
Chisholm, M. H., Errington, R. J., Folting, K. & Huffman, J. C. 1982 *J. Am. chem. Soc.* **104**, 2025–2027.
Corbett, J. D. 1980 *Adv. Chem. Ser.* no. 186, pp. 329–347.
Corbett, J. D. 1981 *a Acct. chem. Res.* **14**, 239–246.
Corbett, J. D. 1981 *b J. solid State Chem.* **39**, 56–74.
Corbett, J. D. 1981 *c J. solid State Chem.* **37**, 335–351.
Cotton, F. A. 1964 *Inorg. Chem.* **3**, 1217.
Cotton, F. A. & Wilkinson, G. 1980 *Advanced inorganic chemistry*, pp. 881–883. New York, Chichester, Brisbane and Toronto: Wiley.
Cotton, F. A. & Fang, A. 1982 *J. Am. chem. Soc.* **104**, 113–119.
Cramer, S. P., Gray, H. B., Dori, Z. & Bino, A. 1979 *J. Am. chem. Soc.* **101**, 2770–2772.
Donohue, P. C. & Katz, L. 1964 *Nature, Lond.* **201**, 180.
Fischer, Ø. 1978 *Appl. Phys.* **16**, 1.
Harned, H. S. 1913 *J. Am. chem. Soc.* **35**, 1078.
Imoto, H. & Simon, A. 1982 *Inorg. Chem.* **21**, 308–319.
Katovic, V. & McCarley, R. E. 1982 (In preparation.)
Kerner-Czeskleba, H. & Tourne, G. 1976 *Bull. Soc. chim. Fr.* no. 729.
Marinder, B.-O. 1977 *Chem. Scr.* **11**, 97–101.
Mattes, R. & Menneman, K. 1977 *Z. anorg. allg. Chem.* **437**, 175–182.
McCarley, R. E., Ryan, T. R. & Torardi, C. C. 1981 *Am. chem. Soc. Symp. Ser.* no. 155, pp. 41–60.
McCarley, R. E. & Lii, K.-H. 1982 (In preparation.)
McCarroll, W. H., Katz, L. & Ward, R. 1957 *J. Am. chem. Soc.* **79**, 5410–5414.
McGinnis, R. N., Ryan, T. R. & McCarley, R. E. 1978 *J. Am. chem. Soc.* **100**, 7900–7902.
Murmann, R. K. & Shelton, M. E. 1980 *J. Am. chem. Soc.*, **102**, 3984–3985.
Perrin, C., Chevrel, R. & Sergent, M. 1975 *C.r. hebd. Séanc. Acad. Sci., Paris* C **281**, 23–25.
Richens, D. T. & Sykes, A. G. 1982 *Inorg. Chem.* **21**, 418–422.
Ryan, T. R. & McCarley, R. E. 1982 *Inorg. Chem.* **21**, 2072–2079.
Schäfer, H. & von Schnering, H.-G. 1964 *Angew. Chem.* **76**, 833–849.
Schäfer, H., von Schnering, H.-G., Niehus, K.-J. & Nieder-Vahrenholz, H. G. 1965 *J. less-commun Metals* **9**, 95–104.
Simon, A., von Schnering, H.-G., Wohrle, H. & Schäfer, H. 1965 *Z. anorg. allg. Chem.* **339**, 155–170.
Simon, A. & von Schnering, H.-G. 1966 *J. less-common Metals* **11**, 31.
Simon, A. 1981 *Angew. Chem. int. Ed Engl.* **20**, 1–22.
Stensvad, S., Helland, B. J., Babich, M. W., Jacobson, R. A. & McCarley, R. E. 1978 *J. Am. chem. Soc.* **100**, 6257–6258.
Torardi, C. C. & McCarley, R. E. 1979 *J. Am. chem. Soc.* **101**, 3963–3964.
Torardi, C. C. & McCarley, R. E. 1981 *J. solid State Chem.* **37**, 393–397.
Torardi, C. C. & McCarley, R. E. 1982 (In preparation.)
Vandenberg, J. M. & Brasen, D. 1975 *J. solid State Chem.* **14**, 203–208.

Phil. Trans. R. Soc. Lond. A **308**, 159–166 (1982)   [159]
Printed in Great Britain

Iron–molybdenum–sulphur clusters

By C. D. Garner, S. R. Acott, G. Christou[†], D. Collison,
F. E. Mabbs and V. Petrouleas[‡]
Department of Chemistry, Manchester University, Manchester M13 9PL, U.K.

The recognition that molybdenum is present in the nitrogenase enzymes as an extractable cofactor, containing iron, molybdenum and sulphur in the ratios 6–8 : 1 : 4–6, has stimulated investigations concerned with the synthesis and characterization of compounds containing these three elements. This paper describes the preparation and structure of complexes containing an Fe–Mo–S framework, most of these systems being classified according to whether they contain a MoS_4 group coordinated as a bidentate ligand to one or two iron atoms, or an Fe_3MoS_4 cubane-like cluster. Several physical properties of these complexes are presented, with reference to the corresponding properties of the iron–molybdenum cofactor of the nitrogenases, especially for those complexes that contain a pair of Fe_3MoS_4 cubane-like clusters.

Introduction

The nitrogenase enzymes occur in both symbiotic and free-living organisms that grow either aerobically or anaerobically. The most studied nitrogenases are those from free-living heterotrophs since these are most readily available and can be cultured on a reasonably large scale in the laboratory. In all cases so far investigated, nitrogenase is constituted of two proteins: a ferredoxin protein, of relative molecular mass *ca.* 60 000 with an $\{Fe_4S_4(Scys)_4\}$ centre, and a molybdoferredoxin protein, of relative molecular mass *ca.* 220 000 with approximately 30–32 Fe, 30–32 S^{2-}, and 2Mo per molecule. The ferredoxin mediates electron transfer to the molybdoferredoxin, this transfer requiring the hydrolysis of ATP, and the latter protein contains the catalytic site(s) for the coupled proton–electron transfers to the enzyme's substrates (Hardy *et al.* 1977; Mortenson & Thorneley 1979; Gibson & Newton 1981). The results of extrusion (Shah & Brill 1977; Kurtz *et al.* 1979) and spectroscopic (Rawlings *et al.* 1978; Huynh *et al.* 1980) studies are consistent with the metal atoms of the molybdoferredoxin porteins occurring as four Fe_4S_4 (or P-type) centres and two iron–molybdenum cofactor (FeMoco), $Fe_{6-8}MoS_{4-6}$ (or M-type) centres; another centre (labelled S), accounting for some 5% of the iron, may also be present. The generally held view is that the FeMoco centre(s) represent the site(s) for substrate reduction, and this view is reinforced by the inactivity of the *Azotobacter vinelandii* UW45 mutant protein that lacks these centres (Shah & Brill 1977).

An important aspect of the characterization of the M-type centre and the *N*-methylformamide-extracted cofactor (FeMoco) has been their e.p.r. spectra, which correspond to an $S = \frac{3}{2}$ system, with principal apparent *g* values of *ca.* 4.6, 3.3 and 2.0. ^{57}Fe Mössbauer spectroscopic studies have shown that the native protein and FeMoco contain one $S = \frac{3}{2}$ centre per molybednum, each

---

† Present address: Department of Chemistry, Imperial College of Science and Technology, South Kensington, London SW7 2AY, U.K.
‡ Present address: Greek Atomic Energy Commission, Nuclear Research Center 'Demokritos', Aghia Paraskevi, Attiki, Greece.

magnetic centre containing about six iron atoms in a spin-coupled structure (Rawlings *et al.*1978). This magnetic centre may correspond to the moiety termed the 'molybdenum–iron cluster', which contains about six iron atoms per molybdenum, isolated from the molybdoferredoxin of *Azotobacter vinelandii* by methyl ethyl ketone extraction, the characteristic $M_{e.p.r.}$ ($S = \frac{3}{2}$) signal being manifest when the cluster is transferred from methyl ethyl ketone to N-methylformamide. This 'molybdenum–iron cluster' seems to have (two) less iron atoms than FeMoco and, unlike the latter, fails to activate the UW45 mutant protein (Shah & Brill 1981). The $S = \frac{3}{2}$ level is an intermediate redox state of the FeMoco and M-type centres; oxidation and reduction lead to e.p.r.-silent states and, in the protein, the reduced state is believed to correspond to the catalytically active condition (Burgess *et al.* 1980).

The measurement and interpretation of the molybdenum K-edge X-ray absorption spectrum of the molybdoferredoxin of *Azotobacter vinelandii* and the N-methylformamide-extracted cofactor (Burgess *et al.* 1981; S. D. Conradson & K. O. Hodgson, personal communication 1982) have provided a more definitive characterization of the molybdenum environment than the initial investigation (Cramer *et al.* 1978), which was so important in stimulating chemical endeavours to prepare iron–molybdenum–sulphur clusters as potential analogues of FeMoco. The latest results are consistent with relatively little perturbation of the cofactor upon extrusion and the extended X-ray absorption fine structure (e.X.a.f.s.) for reduced FeMoco can be interpreted by a molybdenum environment consisting of 3 oxygen (or nitrogen) atoms at 2.10 Å†, 3 sulphur atoms at 2.36 Å, and 2.8 iron atoms at 2.68 Å.

PREPARATION AND STRUCTURE OF Fe–Mo–S COMPLEXES

Complexes that possess an Fe–Mo–S framework are conveniently classified according to whether they involve (1) a tetrathiomolybdate group coordinated to one or two iron atoms, or (2) an Fe_3MoS_4 cubane-like cluster. The majority of the syntheses to the complexes of type 1 (Coucouvanis 1981) have involved a reaction between an iron(II) complex and a $[MoS_4]^{2-}$ salt, whereas the formation of Fe_3MoS_4 cubane-like cluster complexes has been achieved by reacting $FeCl_3$ with a $[MoS_4]^{2-}$ salt, in the presence of sufficient thiolate for ligation and reduction of the metal atoms (Christou & Garner 1980*a*; Holm 1981 and references therein). The only Fe–Mo–S complex currently known that does not fit into this classification is $[\{(C_2H_4S_2)\,Mo(S)\,S_2\}_2Fe]^{3-}$ (Dahlstrom *et al.* 1981).

Each of the synthetic routes currently available for the formation of type 2 complexes yields a dimer, consisting of two $\{(RSFe)_3MoS_4\}$ cubane-like clusters linked across their molybdenum centres by a $(\mu_2\text{-OMe})_3$, as in $[Fe_6Mo_2S_8(SPh)_6(OMe)_3]^{3-}$, a $(\mu_2\text{-SR})_3$, as in $[Fe_6Mo_2S_8(SR)_9]^{3-}$ ($R =$ e.g. Et, CH_2CH_2OH, Ph, $C_6H_4\text{-4-}X$; $X = H$, Me, or Cl) (see figure 1), a $(\mu_2\text{-S})\,(\mu_2\text{-SR})_2$, as in $[Fe_6Mo_2S_9(SEt)_8]^{3-}$, or a $(\mu_2\text{-SR})_3Fe^{II\ or\ III}(\mu_2\text{-SR})_3$ as in $[Fe_7Mo_2S_8(SR)_{12}]^{4-\ or\ 3-}$ ($R =$ e.g. Et or CH_2Ph) arrangement. Some ligand substitution reactions of these clusters have been developed, and the terminal, iron-bound thiolates shown to be readily replaced by other thiols (R′S) or halides ($X =$ Cl or Br), by treatment with R′SH or PhCOX, respectively (Christou & Garner 1980*b*; Palermo *et al.* 1982). The molybdenum centres are kinetically inert to such reactions and it has so far only been possible to achieve substitution thereon via the $(\mu_2\text{-SR})_3$ $Fe(\mu_2\text{-SR})_3$ complexes; reactions with catechols have been developed to produce not only complexes with a single Fe_3MoS_4 cluster but also an important and versatile substitution chemistry at the molybdenum.

† $1\text{Å} = 10^{-10}\,\text{m} = 10^{-1}\,\text{nm}.$

The formation of complexes containing Fe–W–S clusters has proceeded in parallel with the corresponding studies for the Fe–Mo–S clusters. Comparisons between such complexes of these two metals are important because tungsten, although incorporated into the molybdoferredoxin of nitrogenase (Benemann *et al.* 1973), does not lead to a functional enzyme (Shah & Brill 1977, 1981). The structural data obtained for type 1 and type 2 Fe–M–S (M = Mo or W) complexes indicate that the substitution of tungsten for molybdenum produces only a small change in

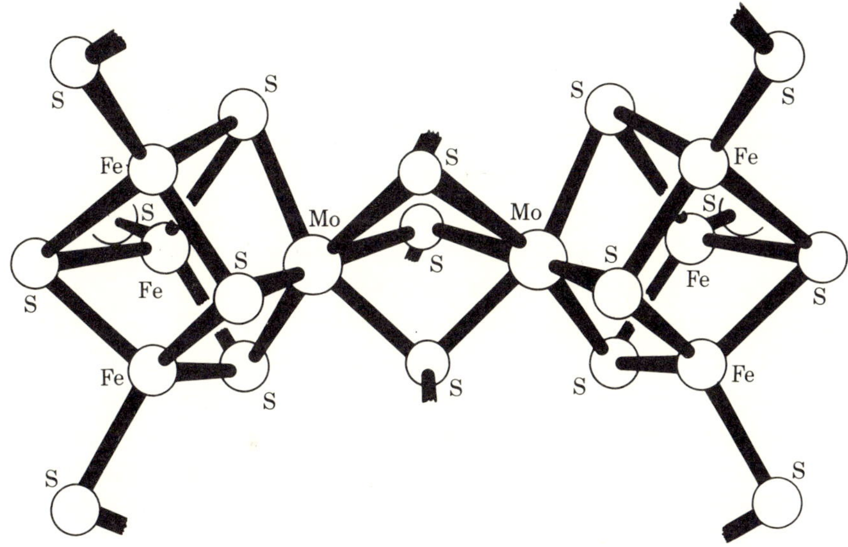

FIGURE 1. Dimeric Fe_3MoS_4 cubane-like framework in $[Fe_6Mo_2S_8(SR)_9]^{3-}$ complexes.

TABLE 1. COMPARISON OF MOLYBDENUM ENVIRONMENTS IN N-METHYLFORMAMIDE-EXTRACTED, $S_2O_4^{2-}$ REDUCED, FeMoco (BURGESS *et al.* 1981; S. D. CONRADSON & K. O. HODGSON, PERSONAL COMMUNICATION 1982) AND $[Fe_6Mo_2S_8(SPh)_9(OMe)_3]^{3-}$

| | FeMoco | | $[Fe_6Mo_2S_8(SPh)_6(OMe)_3]^{3-}$ | |
contact	number	distance/Å	number	distance/Å
Mo–O†	3	2.09	3	2.119 (8)
Mo–S	3	2.36	3	2.343 (3)
Mo–Fe	2.8	2.68	3	2.725 (3)

† Mo–O or Mo–N.

dimensions for type 1 and an insignificant change in dimensions for type 2. The only simple chemical distinction between molybdenum and tungsten to emerge so far is that in Fe_3MS_4 complexes tungsten is a 'harder' metal centre than molybdenum, forming the $(\mu_2$–OMe$)_3$ complex $[Fe_6W_2S_8(SPh)_6(OMe)_3]^{3-}$ under the corresponding conditions that produce the $(\mu_2$–SPh$)_3$ complexes $[Fe_6Mo_2S_8(SPh)_9]^{3-}$ (Christou & Garner 1980a).

The dimensions of the Fe_3MoS_4 cubane-like cluster are essentially independent of the nature of the appended ligands. The number and length of the Mo–S and Mo–Fe contacts within the cubane-like core and, for $[Fe_6Mo_2S_8(SPh)_6(OMe)_3]^{3-}$ (and $[Fe_4MoS_4(SEt)_3(catecholate)_3]^{3-}$), the Mo–O contacts external to the core are in good agreement with the e.X.a.f.s. results of Burgess *et al.* (1981) and S. D. Conradson & K. O. Hodgson, personal communication 1982) (see table 1). The Mo–S and Mo–Fe contacts for the type 1 complexes do not agree with the e.X.a.f.s.

predictions as well as those for type 2; $[Cl_2FeS_2MoS_2FeCl_2]^{2-}$ probably has the closest correspondence observed for a type 1 complex, with 2 Mo–S contacts of 2.204 (5) Å and 2 Mo–Fe of 2.775 (6) Å (Coucouvanis 1981, and references therein).

Electronic structure of Fe_3MoS_4 cubane-like clusters

The total of the formal oxidation states of the metal atoms in the cubane-like cores of the complexes isolated from the $FeCl_3/RS^-/[MoS_4]^{2-}$ reactions is eleven. Thus, the probable description of these oxidation states ranges from $(3Fe^{II}+Mo^V)$ to $(3Fe^{III}+Mo^{II})$. The application of ^{57}Fe Mössbauer spectroscopy (Christou et al. 1980; Holm 1981) has demonstrated that all of the iron atoms are essentially equivalent and have an oxidation state of ca. 2.5. This narrows the choice of formal oxidation state descriptions to $(2Fe^{II}Fe^{III}Mo^{IV})$ or $(Fe^{II}2Fe^{III}Mo^{III})$. ^1H n.m.r. spectroscopy is a useful characterization of these clusters and provides a clear indication that $[Fe_6Mo_2S_8(SR)_9]^{3-}$ and related clusters are paramagnets, with most of the paramagnetism being centred on the iron atoms (Christou & Garner 1980 a).

We have investigated the bulk magnetic susceptibilities and e.p.r. spectra of several of the $[Fe_6M_2S_8L_9]^{3-}$ (M = Mo or W) complexes and interpreted these data in terms of antiferromagnetic spin-coupling between the iron atoms of these complexes; the electrons formally associated with molybdenum (or tungsten) are assumed to be spin-paired, because of electronic exchange interactions across the bridging region or the trigonally distorted octahedral geometry at the metal, or both (Christou et al. 1981, 1982). The exchange interaction between the (high-spin) iron atoms,

$$H_{ex} = -2\alpha J\hat{S}_1 \cdot \hat{S}_2 - 2J\hat{S}^* \cdot \hat{S}_3, \tag{1}$$

within a Fe_3MS_4 cluster is presumed to be the dominant perturbation; centres Fe_1 and Fe_2 are taken to have the same oxidation state and are initially coupled to give a set of resultant spins S^*, which are then coupled with the spin, S_3, of the other iron atom to produce the resultant spin, S', for the individual cube. The inter-cube coupling within an anion (βJ), the applied magnetic field (H) and zero-field splitting (D) have been considered as simultaneous perturbations via the Hamiltonian

$$Hp = \sum_{i=a}^{b} [g_{\parallel}\beta H_z \hat{S}'_z(i) + g_{\perp}\beta\{H_x \hat{S}'_x(i) + H_y \hat{S}'_y(i)\} + D\{\hat{S}'^2_z(i) - \tfrac{1}{3}\hat{S}'(i)(\hat{S}'(i)+1)\}] - 2\beta J\hat{S}'_a \cdot \hat{S}'_b. \tag{2}$$

Equation (2) was applied to product spin functions $|S'(a) M_s(a), S'(b) M_s(b)\rangle$, zero-order energies of which were taken as the sum of the energies of $S'(a)$ and $S'(b)$ resulting from the application of (1). Also, all possible combinations of $S'(a)$ and $S'(b)$ that would lead to thermally occupied states were considered. Magnetic susceptibilities were calculated (Christou et al. 1982) and e.p.r. spectra were simulated (Collison & Mabbs 1982) as described elsewhere. The magnetic properties of $[NEt_4]_3[Fe_6M_2S_8(SPh)_6(OMe)_3]$ (M = Mo or W), $[NBu^n_4]_3[Fe_6Mo_2S_8(SPh)_9]$ and $[NEt_4]_3[Fe_6W_2S_8(SEt)_9]$ has been interpreted by this approach for both the $\{2Fe^{II}Fe^{III}Mo^{IV}\}$ and the $\{Fe^{II}2F^{III}Mo^{III}\}$ models, although rather different values of the parameters, mainly g and α, are required. However, the successful simulation of the solid-state e.p.r. spectra has been achieved only for the $\{2Fe^{II}Fe^{III}Mo^{IV}\}$ model, and then only if a small amount of inter-cube coupling was included. On each cube, only two levels were found to be thermally populated at $\leqslant 10\,K$, one with $S' = \tfrac{1}{2}$ and another $S' = \tfrac{3}{2}$. An interpretation, by using the $\{2Fe^{II}Fe^{III}Mo^{IV}\}$

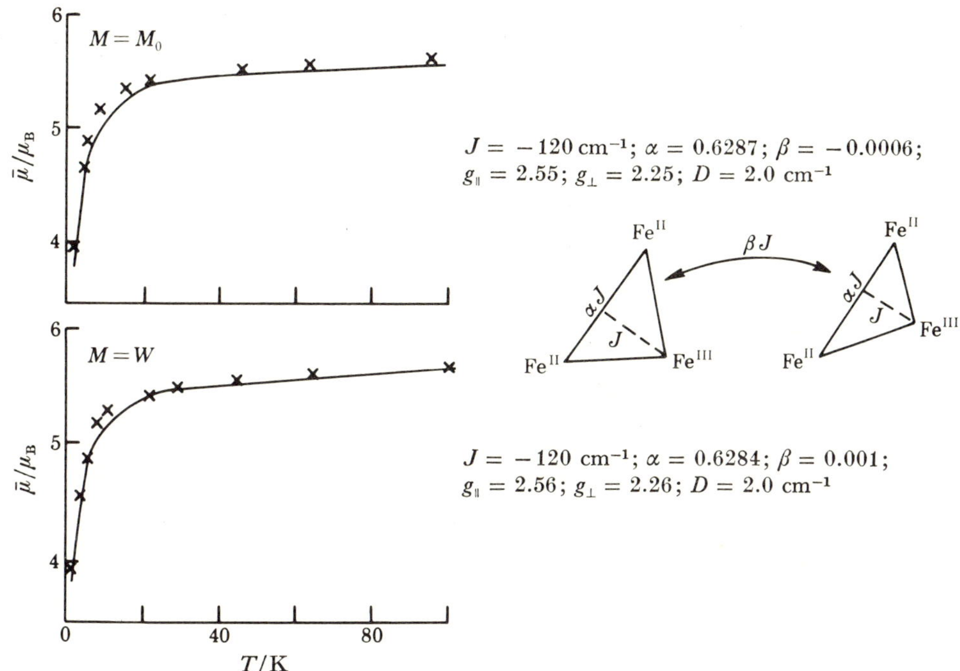

FIGURE 2. Variation of $\bar{\mu}$ with temperature for $[\text{NEt}_4]_3[\text{Fe}_6\text{M}_2\text{S}_8(\text{SPh})_6(\text{OMe})_3]$ (M = Mo or W) and interpretation using an antiferromagnetic spin-coupling model within two high-spin $\{2\text{Fe}^{II}\text{Fe}^{III}\}$ moieties and allowing weak coupling between these groups of iron atoms. ✕, Experiment; —, calculation.

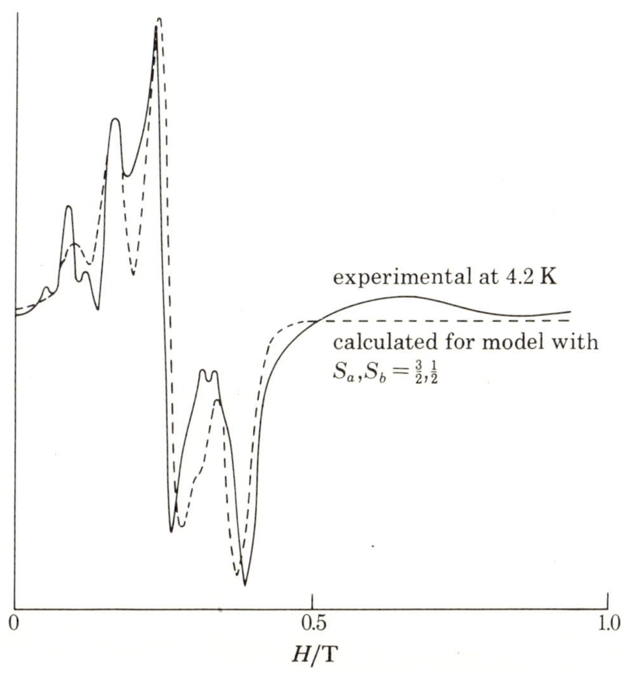

FIGURE 3. E.p.r. spectrum of powdered $[\text{NEt}_4]_3[\text{Fe}_6\text{W}_2\text{S}_8(\text{SPh})_6(\text{OMe})_3]$ at 4.2 K (——) and a simulation (– – –), $\nu = 9.238$ GHz ($S_i\beta JS_j$, where $\beta J = -0.03\ \text{cm}^{-1}$, $E_a(S' = \tfrac{1}{2}) = E_b(S' = \tfrac{1}{2}) = 0.0\ \text{cm}^{-1}$, $E_a(S' = \tfrac{3}{2}) = E_b(S' = \tfrac{3}{2}) = 4.0\ \text{cm}^{-1}$. Line widths for $S'_a = \tfrac{3}{2}$ coupled with $S'_b = \tfrac{1}{2}$ (and vice versa), $w_{\parallel} = 300$ G (30 mT), $w_{\perp} = 450$ G (45 mT); $S'_a = \tfrac{1}{2}$ with $S'_b = \tfrac{1}{2}$, $w_{\parallel} = w_{\perp} = 1000$ G (100 mT); $S'_a = \tfrac{3}{2}$ with $S'_b = \tfrac{3}{2}$, $w_{\parallel} = w_{\perp} = 2000$ G (200 mT); $g_x = g_y = g_z = 2.25$; $D = 2.0\ \text{cm}^{-1}$.

model, of the variation of $\bar{\mu}$ with temperature for the $[NEt_4]_3[Fe_6M_2S_8(SPh)_6(OMe)_3]$ salts is shown in figure 2. For powdered $[NEt_4]_3[Fe_6W_2S_8(SPh)_6(OMe)_3]$ the observed e.p.r. spectrum and a simulated spectrum are shown in figure 3 and the principal features are well reproduced by this model, considering only the coupling of $S'_a = \frac{1}{2}$ with $S'_b = \frac{3}{2}$, and vice versa.

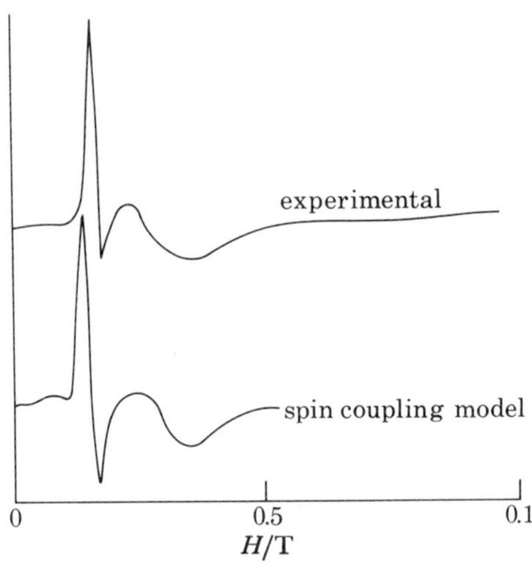

FIGURE 4. E.p.r. spectrum of $[NEt_4]_3[Fe_6W_2S_8(SPh)_6(OMe)_3]$ in MeCN at 10 K, $\nu = 9.238$ GHz; spectrum calculated for $g_\parallel = 2.10$, $g_\perp = 2.20$; $|\beta J| \leqslant 0.012$ cm^{-1}; line widths for $S'_a =$ coupled with $S'_b = \frac{3}{2}$, $w_\parallel = 400$ G (40 mT), $w_\perp = 150$ G (15 mT); $S'_a = \frac{3}{2}$ with $S'_b = \frac{1}{2}$, $w_\parallel = w_\perp = 1400$ G (140 mT). Otherwise, parameters the same as for figure 3.

This simple antiferromagnetic coupling model has therefore allowed a reasonably successful interpretation of the magnetic and e.p.r. properties of these Fe_3MS_4 (M = Mo or W) cubane-like dimers. Although these complexes do not display e.p.r. spectra identical to $M_{e.p.r.}$, we are encouraged by the results obtained, particularly because the appropriate spin states ($S = \frac{3}{2}, \frac{1}{2}$) appear to be involved. The appearance of the e.p.r. spectrum of these complexes can change dramatically from solid state to solution (cf. figures 3 and 4), and this behaviour is attributed to small structural differences perturbing the magnetic couplings or to intermolecular interactions changing the relaxation mechanisms thereby affecting the line widths for the e.p.r. transitions, or both. These observations may be relevant to the significant change in the appearance of the e.p.r. spectrum of the 'molybdenum–iron cluster' of nitrogenase, from methyl ethyl ketone to N-methylformamide solution (Shah & Brill 1981). Finally, we note that the e.p.r. spectrum of these molybdenum-containing complexes involve no observable [95,97]Mo hyperfine couplings and that such couplings are not seen for FeMoco and $M_{e.p.r.}$; for the chemical systems, this observation is consistent with the assumption that the unpaired electron spin density is localized primarily on the iron atoms.

Redox properties

The redox properties of the Fe_3MS_4 (M = Mo or W) cubane-like complexes have been described in detail elsewhere (Christou *et al.* 1980; Acott *et al.* 1982; Holm 1981); the type 1 complexes have a disappointingly restricted redox chemistry. We merely wish to observe here that each Fe_3MS_4 core, in the oxidation level encountered in $[Fe_6M_2S_8L_9]^{3-}$ complexes, is capable of undergoing one reversible and then one irreversible one-electron reduction, and a reversible one-electron oxidation. The dimeric nature of these complexes results in these electron transfers appearing in pairs (see figure 5), producing potentially valuable two-electron transfers. However, as FeMoco appears to be monomeric in molybdenum, the relevance of the chemical studies to the biochemical system lies in the availability of one state more oxidized and one state more reduced than the e.p.r.-active $(S = \frac{3}{2})$ state.

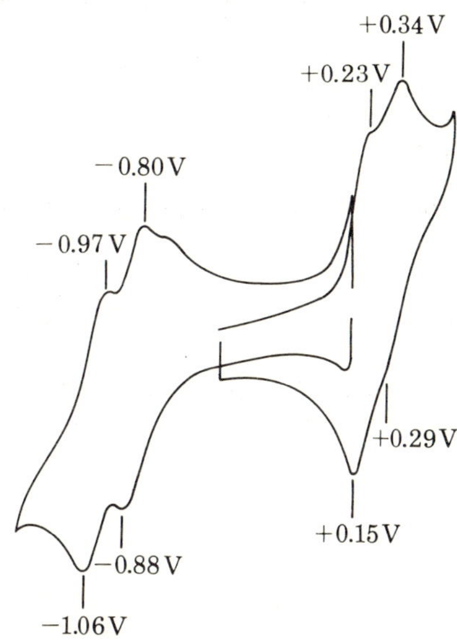

FIGURE 5. Cyclic voltammogram for $[NEt_4]_3[Fe_6Mo_2S_8(SPh)_6(OMe)_3]$ in MeCN (against standard calomel electrode).

CONCLUSIONS

The stimulus of the isolation (Shah & Brill 1977) and e.X.a.f.s. characterization (Cramer *et al.* 1978) of FeMoco has led to the development of an interesting new area of chemistry, that is complexes containing Fe–Mo–S clusters. Of the complex currently characterized, those containing Fe_3MoS_4 cubane-like clusters appear to be the most relevant to FeMoco, in the particular respects of their molybdenum environment, e.p.r. spectra and redox properties. As yet, however, the detailed nature of FeMoco remains obscure, especially in terms of the appended ligands and the location of the iron atoms additional to those within *ca.* 2.7 Å of the molybdenum.

We look forward to further developments in this challenging topic and thank the S.E.R.C. for financial support.

REFERENCES

Acott, S. R., Christou, G., Garner, C. D. & Pickett, C. J. 1982 (Submitted.)
Benemann, J. R., Smith, G. M., Kostel, P. J. & McKenna, C. E. 1973 *FEBS Lett.* **29**, 219–221.
Burgess, B. K., Stiefel, E. I. & Newton, W. E. 1980 *J. biol. Chem.* **255**, 353–356.
Burgess, B. K., Yang, S.-S., You, C.-B., Li, J.-G., Friesen, G. D., Pan, W.-H., Stiefel, E. I., Newton, W. E., Conradson, S. D. & Hodgson, K. O. 1981 In *Current perspectives in nitrogen fixation* (ed. A. H. Gibson & W. E. Newton), pp. 71–74. Canberra: Australian Academy of Science.
Christou, G., Collison, D., Garner, C. D., Acott, S. R., Mabbs, F. E. & Petrouleas, V. 1982 *J. chem. Soc. Dalton Trans.*, pp. 1575–1585.
Christou, G., Collison, D., Garner, C. D., Mabbs, F. E. & Petrouleas, V. 1981 *Inorg. nucl. Chem. Lett.* **17**, 137–140.
Christou, G. & Garner, C. D. 1980*a* *J. chem. Soc. Dalton Trans.*, pp. 2354–2362.

Christou, G. & Garner, C. D. 1980*b* *J. chem. Soc. chem. Commun.*, pp. 613–614.

Christou, G., Garner, C. D., Miller, R. M., Johnson, C. E. & Rush, J. D. 1980 *J. chem. Soc. Dalton Trans.*, pp. 2363–2368.

Collison, D. & Mabbs, F. E. 1982 *J. chem. Soc. Dalton Trans.*, pp. 1565–1574.

Coucouvanis, D. 1981 *Acct. chem. Res.* **14**, 201–209.

Cramer, S. P., Gillum, W. O., Hodgson, K. O., Mortenson, L. E., Stiefel, E. I., Chisnell, J. R., Brill, W. J. & Shah, V. K. 1978 *J. Am. chem. Soc.* **100**, 3814–3819.

Dahlstrom, P. L., Kumar, S. & Zubieta, J. 1981 *J. chem. Soc. chem. Commun.*, pp. 411–412.

Gibson, A. H. & Newton, W. E. (eds) 1981 *Current perspectives in nitrogen fixation.* Canberra: Australian Academy of Science.

Hardy, R. W. F., Bottomly, F. & Burns, R. C. (eds) 1977 *A treatise on dinitrogen fixation*, Sections I and II (*Inorganic and physical chemistry and biochemistry*). New York: Wiley.

Holm, R. H. 1981 *Chem. Soc. Rev.* **10**, 455–490.

Hunyh, B. H., Henzl, M. T., Christner, J. A., Zimmerman, R., Orme-Johnson, W. H. & Münck, E. 1980 *Biochim. biophys. Acta* **623**, 124–138.

Kurtz, Jr., D. M., McMillan, R. S., Burgess, B. K., Mortenson, L. E. & Holm, R. H. 1979 *Proc. natn. Acad. Sci. U.S.A.* **76**, 4986–4989.

Mortenson, L. E. & Thorneley, R. N. F. 1979 *A. Rev. Biochem.* **48**, 387–418.

Palermo, R. E., Power, P. P. & Holm, R. H. 1982 *Inorg. Chem.* **21**, 173–181.

Rawlings, J., Shah, V. K., Chisnell, J. R., Brill, W. J., Zimmerman, R., Münck, E. & Orme-Johnson, W. H. 1978 *J. biol. Chem.* **253**, 1001–1004.

Shah, V. K. & Brill, W. J. 1977 *Proc. natn. Acad. Sci. U.S.A.* **74**, 3249–3253.

Shah, V. K. & Brill, W. J. 1981 *Proc. natn. Acad. Sci. U.S.A.* **78**, 3438–3440.